I0493024

Interdisciplinary Perspectives in Science Fiction

by
Mariella Scerri, Victor Grech and David J. Zammit

for SciFi Malta
www.scifi.ithams.com

and

ITHAMS
The Institute of Technology, Humanities, the Arts, Medicine and Science

With the Collaboration of
HUMS
The Humanities, Medicine and Science Programme at the University of Malta

Science Fiction Symposium, Malta, 17 July 2015
http://www.scifisymposium.com

Copyright © 2013: The Organisers, SciFi Malta and ITHAMS

All rights reserved. No part of this publication may be reproduced, distributed, or transmitted in any form or by any means, including photocopying, recording, or other electronic or mechanical methods, without the prior written permission of the publisher, except in the case of brief quotations embodied in critical reviews and certain other non-commercial uses permitted by copyright law. For permission requests, write to the publisher, addressed "Attention: Permissions Coordinator," at

The organisers
Institute of Technology, Humanities, the Arts, Medicine and Science
Faculty of Science Fiction
Sir Temi Zammit Buildings, LS1.0.1.2,
Life Sciences Park
San Gwann SGN3000, Malta
Email: info@ithams.com / info@scifi-malta.com
Web: www.scifi.ithams.com

Table of Contents

Acknowledgements

We wish to thank all friends and colleague who contributed to the long process of making this book possible. We particularly wish to thank the Rector of the University of Malta, Prof Juanito Camilleri for his active support within the HUMS programme, the Dean of the Faculty of Arts, Prof Domnic Fenech who hosted the Science Fiction Symposium at the Faculty of Arts Library, as well as Prof Joseph Cacciattolo, co-ordinator of HUMS for his support. Special thanks go to Prof Ivan Callus for his forbearance with the writers, for his invaluable help and input and his constant encouragement. Without his help this book would not have been possible. We also thank the Institute of Technology, Humanities, the Arts, Medicine and Science, represented by Mr Charles Busuttil, executive secretary, who greatly helped in the organisation of the symposium. We would like to thank Shadow Services Group, especially Daryl Zammit, for ICT help and support. We cannot also but thank our families and close friends for their forbearance in this project and lively support and encouragement during this lengthy process.

Foreword

Interdisciplinary perspectives in Science Fiction is the second book in what is to become a series of books created and published to explore the interface of medicine, science and humanities in relation to the genre of science fiction. The collection of papers are based on oral presentations delivered during a Science Fiction Symposium held in Malta in July, 2015 from various researchers working in different fields. A similar book on Star Trek: Interdisciplinary Perspectives in Theory and Practice was published last year after the first academic symposium on Star Trek was held in Malta in July, 2014. The writers felt the need to document the interesting work delivered in both symposia, hence these works.

Preparations for another Star Trek Symposium to celebrate the 50th anniversary from the launch of Star Trek in 1966 are also under way, creating a niche for a third publication. The enthusiasm, strong will and dedication of the writers, as well as the delegates' motivation and encouragement make all this possible, giving them the impetus to actively engage themselves in further commitments of this type.

Science Fiction (SF) is difficult to define. Perhaps the most accurate description was accorded by the critic Darko Suvin who opined that the genre "is distinguished by the narrative dominance of a fictional novelty (novum, innovation) validated both by being continuous with a body of already existing cognitions and by being [...] based on cognitive logic."

Based on this definition, the organisers assembled an academic meeting devoted to SF, held in Malta at the Faculty of Arts Library, University of Malta on the 17 July 2015. The meeting was organised by Prof Victor Grech, Ms Mariella Scerri and Dr David Zammit with the help of Malta Institute for Medical Education. The event was held under the aegis of HUMS, The Humanities, Medicine and Sciences Programme at the University of Malta. David Zammit's company (Shadowservices) created the event's website and acted as web host.

The meeting differed from the traditional fan-based convention in that it served as an international platform for academics from across many disciplines to meet and explore the intersection of humanities and sciences through SF. A variety of interesting papers were presented, some jointly, as will be shown in this volume.

The event was opened by Her Excellency Gina Abercrombie-Winstanley, U.S. Ambassador to the Republic of Malta, who is becoming a regular in these events. Her introductory speech on disability in Star Trek was both inspiring and thought provoking.

The event attracted both local and international delegates, two of which (Chris Pak and Stefan Rabitsch) gave their online presentation via Skype. The meeting was a success, with animated discussions after every presentation. It was also felt that the event should be documented in a volume of papers, hence this second volume of work. Some of these papers were destined to be published elsewhere and hence could not be reproduced here in. Chris Pak's 'Drilling, the Extractive Industries and Sustainability in Science Fiction was an interesting presentation. A 21st century update on 'Alien Conspiracy Theories' was given by Duncan Ayers. In his presentation he illustrated how the stories and postulations of intelligent alien life forms who made their presence felt within human society, have long been described and quoted across centuries. However, he noted that there have never been any formal confirmation of these alien sightings as events of first contact by governmental authorities of any country as yet. This brief update also highlighted the major conspiracy theories regarding alien sightings that have occurred mainly in the last fifteen years and discussed possible reasons for the authorities to prevent such information from leaking to the general public on a global scale.

Another presentation given by Janet Mifsud dealt with neoplatonistic concepts in science fiction in a paper 'Clarke, Q and Avatar – Neoplatonistic Concepts in SciFi?' Metaphysical concepts have been described as the science of first causes, the science of things in so far as they exist or the science of theological sciences. Mifsud claimed that the interaction between cosmology, theology (the science of God) and ontology (the science of being) is reflected in numerous ways in many science fiction narratives. The neoplatonistic concepts of a perfection which is beyond understanding are frequently found in scifi narratives. This paper gave an overview of some examples. With respect to the ending in his 2001: Space Odyssey (1968), Arthur C Clarke remarked, "If this film can be completely understood, then we will have failed." Q, the Star Trek character who appears in Star Trek: The Next Generation, Star Trek: Deep Space Nine and Star Trek: Voyager, is said to be omnipotent. His home, the Q Continuum, is accessible only to the Q and their guests, and the true nature of it is said to be beyond the comprehension of "lesser beings" such as humans so it is shown to humans only in ways they can understand. In Sodenbergh's Solaris (2002), the planet Solaris becomes an artistic representation of the true Unknown, and the unknown is that state in which we all exist in, but in which we create forms of meaning to encapsulate the mystery of this moment within the illusion of the Known. This paper concluded with a discussion on how these concepts resonate with today's audiences living in a highly networked reality, who understand themselves to be individuals, yet at the same time part of a vast ubiquitous networked existence.

We feel that it is important to emphasise that this volume is deliberately crafted to be accessible to academics in all camps and fields, as well as to anyone generally interested in science fiction. It should therefore appeal to academics and science fiction fans alike. Please enjoy! And as the Star Trek adage goes… Live long and prosper!

Mariella Scerri, Victor Grech and David Zammit

Introduction

Her Excellency, United States Ambassador to Malta, Gina K. Abercrombie-Winstanley.

I'm delighted to be here again this year. As you well know, I am an avid fan of Star Trek.
I grew up watching the original – Kirk, Spock, Bones – and I continue to be a fan of the spin-offs and movies.

One cannot discuss interdisciplinary perspectives in science fiction without mentioning 'disability.' Given that 2015 marks the 25th anniversary of the Americans with Disabilities Act, the topic is timely. President Obama is committed to nurturing a society that values the contributions of all of our citizens and residents, including the approximately 50 million people in the U.S. living with disabilities.

President Obama believes that disability rights are not just civil rights to be enforced at home; they are universal rights to be recognized and promoted around the world. Today, 650 million people – 10 percent of the world's population – live with a disability.

Star Trek is also an advocate for disability rights. It has frequently has depicted many aspects of disability, chronic illness and rehabilitation. The narratives often portrayed these individuals in a disability-blind manner, as simply different, like someone with different hair or eye colour. Their environment had been perfectly adapted to optimize their skills and abilities. There were times when the disability became an advantage which helped save the crew.

Disability in Star Trek becomes "differently abled." The show offered a message of hope and a glimpse of how we could remake our world. For example, Dr. Miranda Jones was blind but had exceptional sensory abilities and was able to determine a person's heart rate. Lieutenant Commander Geordi La Forge was also blind and his visor allowed him to see more than other humans. The show bucked convention in portraying them as highly capable officers, regardless of sight.

Ensign Melora Pazlar was unable to walk unassisted because of an evolutionary adaptation to a planet with very low gravity. The crew set up ramps for her to use. Once her basic needs were met, Melora refused any special assistance. She insisted she was not sick and therefore, not in need of curing. Melora wanted her companions to accept her differences and create an environment where she could function independently. Self-reliance is a recurring appeal of people with disabilities.

Science and innovation are key elements in our efforts to create an environment which empowers all people. We design and build much of our environment. Why would we purposefully reject or isolate any community members?

Mediator Riva was deaf yet able to communicate through Chorus, a group of three people who read his thoughts and put them into words to lead peace talks between warring tribes.

It is a concrete example of working together to achieve more by drawing upon his or her special abilities. Eventually, Riva vows to overcome tribal strife by bringing them together to learn his sign language.

In his study, Terry L. Shepherd found that special education teachers can use guided viewing of Star Trek episodes to facilitate discussions of disability issues. This video therapy can help students with disabilities find resolutions to their own feelings and problems through the experiences of Star Trek characters. It can also be used with students without disabilities to promote awareness, understanding, and acceptance of individuals with disabilities.

ENABLE is the key to addressing disability issues. As Star Trek did, we must remove attitudinal and environmental barriers which hinder full participation in society on an equal basis. We must empower and enable every individual to reach his or her full potential. On the International Day of Persons with Disabilities, President Obama reaffirmed "the fundamental principle that those with disabilities are entitled to the same rights and freedoms as everyone else: to belong and fully participate in society, to live with respect and free from discrimination, and to make of their lives what they will."

I hope you will join me in working to bring this aspect of science fiction to today's world.

Thank you and I hope you enjoy the symposium.

Chapter 1.   Contributors

1.1   Ivan Callus

Ivan Callus is Associate Professor in the Department of English at the University of Malta, where he teaches courses in contemporary fiction and literary theory and where he served as Head of Department between 2006 and 2014. He obtained his PhD at the Centre for Critical and Cultural Theory at Cardiff University in 1998, following research on the cahiers d'anagrammes and the unpublished writings of Ferdinand de Saussure, aspects of which have appeared in journals since and which inform a monograph due for publication in 2016.

He is the co-editor of CounterText, a journal published by Edinburgh University Press, and of Discipline and Practice: The (Ir)resistibility of Theory (Bucknell University Press, 2004), Post-Theory/Culture/Criticism (Rodopi, 2004), Cy-Borges: Memories of Posthumanism in the Work of Jorge Luis Borges (Bucknell University Press, 2009), Posthumanist Shakespeares (Palgrave Macmillan, 2012) and Style in Theory: Between Literature and Philosophy (Bloomsbury Academic, 2013). He is the author of numerous journal articles and book chapters on contemporary narrative, poststructuralist literary theory and posthumanism (including work which has appeared or is forthcoming in EJES, Angelaki, Subjectivity, ebr, Forum for Modern Language Studies, Cahiers

Ferdinand de Saussure, Word and Text, Parallax, Comparative Critical Studies and Arcadia). With Stefan Herbrechter he is the editor of the Critical Posthumanisms series (Brill). His current research centres on posthumanism, electronic literature and 21st-century narrative.

1.2 Victor Grech
See about the authors

1.3 Stefan Rabitsch

Stefan 'Steve' Rabitsch is currently a visiting postdoc fellow in American Studies at Alpen-Adria Universität Klagenfurt (AAU) in Austria. He also serves on the board of the Austrian Association for American Studies (AAAS). His research and his classes are dominated by American Culture Studies together with a pronounced focus on Science Fiction Studies across media. He is particularly interested in the discourses and semiotics of historiography and worldbuilding in television, film and video games. He is appreciative of the label 'Academic Trekkie' which is occasionally attached to his name. His dissertation on the secret British history of Star Trek is forthcoming as a monograph at McFarland.

1.4 Mariella Scerri

See about the authors

1.5   Clare Vassallo

Clare E. Vassallo graduated from the University of Malta with B.A. (Gen.) in Philosophy and English Literature with Linguistics, and a B.A. (Hons.) in English with Linguistics. She pursued her interest in the interface between these three areas at the Istituto di Comunicazione, University of Bologna, Italy where she obtained her Ph.D in Semiotics under the tutorship of Prof. Umberto Eco. The emphasis of her research was in the field of Semiotics as literary and cultural theory and as theory of knowledge.

She is currently Associate Professor in the Department of Translation Studies at the University of Malta. She teaches postgraduate courses in the Department of Translation Studies and in the MA Program on Popular Culture and Literary Tradition, as well as undergraduate courses in the English Department.

She is Co-Chair on HUMS – the University of Malta's Humanities, Medicine and Science Program, she is a Board Member of the ESU – the English Speaking Union in Malta, and was previously a Board Member and Chair on the BCA – British Culture Association. She was Consultant to the Ministry of Culture (199-2001). She is currently a Board Member of the NGO on public policy Think Tank TPPI – Today's Public Policy Institute, and has been involved as a Thematic Coordinator on V18 – Valletta European City of Culture 2018. She has served as Judge on a number of national prize events, including the National Book Prize (5 years), the Institute for Journalism Prize (3 years), and others.

1.6 Stefan N. Vella

Dr.Stefan Nicholas Vella is a practicing Lawyer with a Magister Iuris in European and Comparative Law from the University of Malta , and was called at the Bar in 2001. He specialized in Maltese and EC anti-trust law and European Union Law for fourteen years and currently is working with the Ministry for Social Dialogue Consumer Affairs and Civil Liberties as a Senior Legal Officer. His research interests are legal theory, and anti-trust, European Union Law, philosophy, particularly political theory, as well as, theology, especially political theology, and comparative theology.

His understanding of philosophy of law, and the general theorizing of law and its application, helps in his understanding of how legal theory manifests itself in science fiction, as well as the relationship between legal theory and science fiction. Moreover, his interest in political theory helps him analyse how science fiction through speculative literature, defines totalitarianism and the issue of legitimacy of law, and the manifestation of the existence of law through coercion and also how science fiction deals with the different models of the state, notably that of totalitarianism, and the concept of the leviathan state. Currently, he is being actively engaged in the issues of data protection and surveillance, and human rights doctrine relating to administrative procedures, particularly in issues related to anti-trust investigative procedures, as well as, the legislation of measures related to surveillance and the striking of the balance between the right of the individual to data privacy on the one hand, and the right of the state to defend itself, and the duty to provide security to its citizens.

1.7 David Zammit

See about the authors

Chapter 2: Postcolonialism and Science Fiction – Ivan Callus

Science fiction can seem to countenance colonialist gestures. That most famous of lines, "To boldly go where no one has gone before," is emblematic in this regard. In effect, its use of the infinitive is cast as a colonializing imperative, the objectionable potentialities of which are kept at bay by the implicit allure of exploration. "To boldly go …": as if the injunction to make the Empire's reach progressively more interstellar is some kind of overriding, all-consuming objective in the cosmic "To Do" list of humanity entire. It is, in effect, not far removed from Conrad's arresting image in the first chapter of *Heart of Darkness*:

> For a time I would feel I belonged still to a world of straightforward facts; but the feeling would not last long. Something would turn up to scare it away. Once, I remember, we came upon a man-of-war anchored off the coast. There wasn't even a shed there, and she was shelling the bush. It appears the French had one of their wars going on thereabouts. Her ensign dropped limp like a rag; the muzzles of the long six-inch guns stuck out all over the low hull; the greasy, slimy swell swung her up lazily and let her down, swaying her thin masts. In the empty immensity of earth,

sky, and water, *there she was, incomprehensible, firing into a continent*. Pop, would go one of the six-inch guns; a small flame would dart and vanish, a little white smoke would disappear, a tiny projectile would give a feeble screech—and nothing happened. Nothing could happen. There was a touch of insanity in the proceeding, a sense of lugubrious drollery in the sight; and it was not dissipated by somebody on board assuring me earnestly there was a camp of natives—he called them enemies!—hidden out of sight somewhere. (Conrad 115; emphasis added)

The truth staged by this passage is that before the Empire wrote back it was firing back, as well it might (Ashcroft et al.). Colonialist expansion is not, after all, "straightforward," neither then nor in intergalactic futures. Across the *Star Trek* franchise, for instance, the Star Fleet's propensity to have one of their wars thereabouts in deep space means that the action is even more dispersed into the empty immensity of earth, sky, and water. It means that the *Enterprise* and its analogues across science-fiction narrative fire not into an empty continent but into the absolute void itself. It is indeed, incomprehensible: a proceeding with a touch of insanity about it, lugubriously droll. But menacing too, because the "enemies," the other in the void, will turn out to be not "a camp of natives" but an alterity that goes unanticipated and which can surprise. The Empire, in fact, can itself come back at you at more than the speed of light. It does so in K. W. Jeter's *Morlock Night*, to cite an early steampunk novel where the Morlocks hitch the time machine back to Earth in a rewriting of Wells's *The Time Machine*. It can itself boldly *come* where no alien had been before, at least as far as we know – just as occurs, elsewhere, in Wells's *The War of the Worlds*. So that it is with the relief of more than schadenfreude that humanity in this novel sees the alien devastated, undone by scant defence against counter-invasion of its biosystem by earthly microbes: a doubled example of interstellar inter-species colonization, as it happens. The problem, however, is that aliens, like humans, are not above recidivism. They can come back, to (ex)terminate. As the tagline of *Independence Day: Resurgence* has it: "We had twenty years to prepare. So did they."

Alien predation rendered terminally, or renewably, is however not quite the focus here. Rather, another (post)colonialism-informed narrative of Conrad's, *Nostromo*, offers further prompts to this set of brief reflections on postcolonialism and science fiction. *Nostromo* is set in Costaguana, an imaginary South American country, and more specifically in Sulaco, a town as vital to the narrative's atmosphere as Chandrapore turns out to be in Forster's *A Passage to India*, for it is close to a silver-mining concession whose economic significance shapes the avarice-driven events that Conrad recounts. There are two points worth remarking here. The first concerns the well-known episode in the novel where Nostromo, an Italian sailor who has made his living in Costaguana as a formidably indispensable factotum for whoever engages him toward economic ventures of some moment, finds himself with another character, Decoud, in a silver-laden lighter that finds itself adrift in the pitchest of darkness. The description is masterly and chilling, providing not only insight into the psychology of disorientation but also an allegory of postcolonialism's macabrely becalmed plundering – in prefiguration of science fiction narratives, as in *Gravity*, that turn on the fear of losing oneself in orbit, untethered from the mother ship. The colonialist sensibility drawn against the comeuppance of its bold goings is what is portrayed there. The second point is that Nostromo is called, precisely, *nostromo*: an old moniker for sailors, the Italian *nostro uomo* shortened but retaining the sense of "our man," our man out there, as in "Our Man in Havana," to cite a phrase that provides a famous title for Graham Greene but which also stages, implicitly, the identification with our kind, the recognition of exclusionary affinities that motivate, in the end and (un)consciously, the alterity-occluding energies of the colonializing dynamic. The intuition that Forster had at the end of *A Passage to India*, another well-known passage in these contexts, is related: sides are chosen, and in the (post)colonial relation the opting can be elemental and immemorial, and its choices definitive and implacable:

> "Why can't we be friends now?" said the other, holding him affectionately. "It's what I want. It's what you want."
> But the horses didn't want it — they swerved apart; the earth didn't want it, sending up which riders must pass single-file; the temples, the tank, the jail, the palace, the birds, the carrion, the Guest House, that came into view as they issued from the Gap and saw Mau beneath: they didn't want it, they said in their hundred voices, "No, not yet," and the sky said, "No, not there." (Forster 189)

I mention this because it is for nostri uomini that in science fiction we are primed to root (the winsomeness of beings like Spielberg's E.T. notwithstanding). You might be as postcolonially reconstructed as anything, but in The Matrix it is with Neo and his meatworldly band that you are going to identify, not with the Machines or Agent Smith. Science fiction tends to bring us out cheerleading our colonising kind, lest we be colonised instead. Alien chic, to use Neil Badmington's phrase, may be all very well, but we don't want it near us. Human specificity pointedly corrals itself away from the alterity of the alien, the cyborg, or the monster, even as we might encroach upon their territories in those moves of disidentifying appropriation that make for the very physis of the colonial. To use a title from Haraway (2008), when species meet the encounter is unequal, the scope for reciprocity wasted.

It is because reciprocity fails that cyborg manifestoes (Haraway 1990) acquire pertinence: thence posthumanism, a paradigm predicated on technoscientific primacies. As N. Katherine Hayles, in a foundational passage, has it,

> First, the posthuman view privileges informational pattern over material instantiation, so that embodiment in a biological substrate is seen as an accident of history rather than an inevitability of life. Second, the posthuman considers consciousness [...] as an evolutionary upstart trying to claim that it is the whole show when in actuality it is only a minor sideshow. Third, the posthuman view thinks of the body as the original prosthesis we all learn to manipulate, so that extending or replacing the body with other prostheses becomes a continuation of a process that began before we were born. Fourth, and most important, by these and other means, the posthuman view configures

the human being so that it can be seamlessly articulated with intelligent machines. In the posthuman, there are no essential differences or absolute demarcations between bodily existence and computer simulation, cybernetic mechanism and biological organism, robot teleology and human goals. (Hayles 2–3)

There is much here that calls for comment, but a simple and sharp enough point (or, perhaps, provocation) suffices. Posthumanism and science fiction are a discourse of the West. Theirs is a First World imaginary.

Let us consider this. If it is true that posthumanism is, essentially, the study of humanity's precariousness but also the study of humanity's alternative futures, whether imminent or fantastical, then there is some viability in thinking of science fiction as being, to all intents and purposes, the new realism. But even before science fiction in the present helps to fictionally explore what posthumanism verisimilarly opens onto, there is a prior and enabling techno-economic determinism to its imaginary. There is much both in science fiction and in posthumanism that is predicated on First-World-enshrining science, on research into robotics, on Artificial Intelligence and Artificial Life, on the world of particle colliders, on spacecraft-engineering corporations, to name but a few affordances that have, in effect, a regionalism to their affordability. When the Singularity occurs (Kurzweil 2006), the presumption is that it will transcend humanity entire, without distinction of gender, race, or creed. The suspicion however lingers that income and privilege might make the post-Singularity settlement somewhat more alright for some, even then.

A question to ask, therefore, is how do literature and fiction that are not entirely aligned with First World imaginaries represent posthumanist conceits and science fictional scenarios? More specifically: what would a postcolonialist science fiction – as in the Empire writing back science fiction, reimagining it, reconfiguring it – be? For, as will have been grasped, the matter is a little more pressing than can be assuaged with the expedient of the flight deck on the Starship *Enterprise* being diversely open to ethnicities, Vulcans, and androids.

A BBC World Service programme first broadcast on 17 June 2012 provides some perspectives on this. "Is Science Fiction Coming to Africa?" had science fiction writer Lauren Beukes (prizewinning author of *Zoo City,* among other works) in conversation with film directors Neill Blonkamp and Wanuri Kahiu (directors of *District 9* and *Pumzi* respectively), the blogger Jonathan Dotse, who writes (on) African cyberpunk and science fiction, and Nnedi Okorafor, whose prose takes in speculative fiction and fantasy. 'Africa tends to suffer in science fiction,' notes Beukes. "Philip K. Dick wrote the continent off the map in *The Man in the High Castle*," she observes, before pointedly asking, "Is there such a thing as African science fiction?" Her question arises because science fiction is typically associated with "advanced technology," whereas "people struggle to reconcile that with the idea of Africa as underdeveloped, primitive, and Third World." The other participants in the conversation acknowledge that science fiction "is seen as a Western genre." Indeed, "it's *alien*." "Afrofuturism" is, in that context, niche at best, with some risk that attempts at an African science fiction might be "patronizing." But in the same way that *District 9* was able to depict "robots in an alien apartheid," African science fiction can "extrapolate from where we are now" and discover that it both "resonates globally" and in streets across the continent. For instance *Pumzi*, with its depiction of "water wars," was able to depict how "air and water can quickly become a commodity" in a way that reimagines all too immediate situations without risking "issue fatigue." Science can be empowering in Africa, and "one of the ways you can promote science and technology is by writing science fiction and giving people a vision" of their world and their society, "because if science fiction has a purpose, it's a way to refract the real world" (Kahiu 2009).

Clearly then, science fiction is important in critical studies minded to postcolonial issues, because it enables a different outlook onto matters that range from how it is that the Empire writes back in 21st-century technoscientific society to the refraction of posthumanism's conceits. In question here is not advocacy of regionalism within science fiction, but awareness of certain political dimensions to the genre that can otherwise be overlooked. In other words, science fiction – like posthumanism – can seem to represent reaches of genre fiction and/or scholarly discourse that could seem to be not universally writeable, not open to equitable inscription within literature's institutionality. It is enough to make us ask whether literature, which we like to think is not foreclosed to the postcolonial, quite goes far enough in terms of its reconstructive impetus, or whether it is, in fact, a compensation or an inadequacy. For postcolonialist writing, the more impactful challenge arguably lies not with its visibility within literature and its traditions, an objective which has largely been met, but with its approach upon genre fiction and futurism, a wager whose stakes are still in the process of being understood.

References
Cinematography

District 9. Dir. Neill Blomkamp. South Africa: TriStar Pictures, 2009. Film.

E.T. the Extra-Terrestrial. Dir. Steven Spielberg. US: Universal Pictures, 1989. Film.

Gravity. Dir. Alfonso Cuarón. US: Warner Bros, 2013. Film.

Independence Day: Resurgence. Dir. Roland Emmerich. US: Fox, 2016. Film.

Pumzi. Dir. Wanuri Kahiu. Kenya: Focus Features, 2009. Film.

The Matrix. Dir. the Wachowskis. US: Warner Bros, 1999. Film.

Primary Texts

Ashcroft, Bill, Griffiths, Gareth and Tiffin, Helen (eds.). *The Empire Writes Back: Theory and Practice in Post-Colonial Literatures*. New York: Routledge, 1989. Print.

Badmington, Neil. *Alien Chic: Posthumanism and the Other Within.* London: Routledge, 2004. Print.

Beukes, Lauren. *Zoo City*. Johannesburg: Jacana Media, 2010. Print.

Beukes, Lauren. "Is Science Fiction Coming to Africa?" In conversation with Neill Blonkamp, Wanuri Kahiu, Jonathan Dotse, Nnedi Okorafor. BBC World Service, London. 2012. Radio.

Conrad, Joseph. *Heart of Darkness and Other Tales*. Cedric Watts (ed.). Revised edition. Oxford and New York. Oxford University Press, 2008. Print.

Conrad, Joseph. *Nostromo*. Jacques Berthoud and Mara Kalnins (eds.). Oxford and New York: Oxford University Press, 2009. Print.

Dick, Philip K. *The Man in the High Castle*. New York: Putnam, 1962. Print.

Forster, E. M. *A Passage to India*. Oliver Stallybrass (ed.). Harmondsworth: Penguin, 1985. Print.

Greene, Graham. *Our Man in Havana*. London: Vintage, 2001. Print.

Haraway, Donna. *Simians, Cyborgs, and Women: The Reinvention of Nature*. New York: Routledge, 1990. Print.

Haraway, Donna. *When Species Meet*. Minneapolis: University of Minnesota Press, 2008. Print.

Hayles, N. Katherine. *When We Became Posthuman: Virtual Bodies in Cybernetics, Literature, and Informatics*. Chicago: University of Chicago Press, 1999. Print.

Jeter, K. W. *Morlock Night*. Nottingham: Angry Robot, 1979. Print.

Kurzweil, Ray. The Singularity is Near: When Humans Transcend Biology, 2005. Print.

Wells, H. G. *The Time Machine*. Ed. Patrick Parrinder. London: Penguin, 2005. Print.

Wells, H. G. *The War of the Worlds*. Ed. Patrick Parrinder. London: Penguin, 2005. Print.

Chapter 3. The Elicitation of Jung's Shadow in Star Trek[1] – Victor Grech

Introduction

This paper will briefly explore Carl Jung's concept of the shadow, a subconscious archetype, and will scrutinise the myriad ways in which the shadow has manifested in *Star Trek* (ST), all the while probing the ensuing repercussions that these narratives have sought to explore.

Sigmund Freud divided the self into the conscious and the unconscious mind, and the latter was further divided into id (instincts and drive) and superego (conscience). The unconscious mind is not usually accessible to the conscious mind and includes repressed feelings, phobias, desires, traumatic memories and emotions. These are socially unacceptable and the individual is therefore actively averse to acknowledging them, hence their suppression.

Carl Jung further developed this notion by dividing the unconscious into a personal and a collective unconscious. The former resembles the Freudian concept of the unconscious while the latter comprises

---

[1] This piece was originally published in the NYRSF Review as Grech, Victor. "The Elicitation of Jung's Shadow in *Star Trek*" *The New York Review of Science Fiction.* 306 (2014): 1, 14-20.

inherited psychic structures and archetypes that are shared by the entire race. Archetypes are universal templates that embrace common classes of memories and interpretations and may be used to interpret behaviours. Jung delineated five major archetypes within the individual:

1. The Self, the control centre.
2. The Shadow, which contains objects with which the ego does not consciously or readily identify.
3. The Anima, the feminine image in a man's psyche, or the Animus, the masculine image in a woman's psyche.
4. The Persona, the mask which the individual presents to the world.

This concept is tangentially alluded to in ST, when Captain Picard states that the extinct "Kurlan civilisation believed that an individual was a community of individuals. Inside us are many voices, each with its own desires, its own style, its own view of the world" (Frakes, "The Chase").

Since one is likelier to repress one's least desirable personality traits, the shadow is largely negative and comprises "an aspect (...) of the individual's personality which is objectified or personified through projection" (Woods and Harmon 170). In narratives, "[t]hese contents or elements of the unconsciousness are manifested as archetypes, which in turn are expressed symbolically" (Woods and Harmon 171), such that the reader or viewer may readily comprehend the nature of the shadow which is reified as an actual person.

Individuation is the process whereby components such as personal experiences and archetypes are merged and integrated, producing a stable, functioning and balanced individual. This includes the shadow since the individual "constantly needs the renewal that begins with a descent into his own darkness" (Jung, "Mysterium" 334), a repellent, albeit necessary undertaking.

This does, however, carry the risk of a shadow takeover, since Jung believed that this archetype is the strongest of all, in Jekyll and Hyde fashion, since the "acknowledgement of the shadow must be a continuous process throughout one's life" (Hart 92). Jung also believed that the excess of any force, including the shadow, inevitably results in its opposite coming into being, with an equilibrium reached in a *coincidentiaoppositorum*, such that the shadow can be isolated, studied, accepted and the process of individuation embarked upon. Indeed, this Heraclitean concern with opposites and their interactions in the formation of the structure and functioning of the human psyche was a prime tenet in Jung's work.

The physical manifestation of the shadow is extremely common in ST as it allows the psychological exploration of the human condition through "an explicit dialogue or interplay between the parts of the personality that coexist uneasily within each of us" (Lundeen and Wagner 74). The science fiction genre and ST itself are highly suited to such explorations "given the special premises of its science fictionworld (to say nothing of cinematic special effects), the Doppelganger can so easily be made incarnate (Lundeen and Wagner 74).

Arguably, "[t]he breadth and depth of Star Trek's appeal can (…) more easily be understood by referring to basic and universal psychic structures" (Blair 311). Indeed, "[t]he simultaneous presence on screen of two identical or nearly identical characters played by the same actor is so common in Star Trek that one might consider it a stock Trek device" (Lundeen and Wagner 71). This paper will catalogue and categorise all manifestations of the shadow in ST and any relevant references thereto, in the same way that Roger Robert's seminal work *A Psychoanalytic Study of the Double in Literature* (1970) studied more classical narratives.

The transporter offers a convenient medium whereby the individual may be deliberately or accidentally split, and the shadow exposed (Grech, "The Trick"). As observed by Spock, this provides "an unusual opportunity to appraise the human mind, or to examine, in Earth terms, the roles of good and evil in a man" (Penn, "The Enemy Within"). However, other opportunities arise that do not include ST's transporter and a variety of these occurrences will be also explored, under the categories of shadows that must be physically integrated, shadows that cannot be reintegrated, unleashment of the shadow by the acquisition of excessive powers, racial collective shadows and the actual subtraction of the shadow. For the purposes of this essay, the terms "fragmentation,""doubling," and "decomposition" will be used synonymously. (Rogers 4-5).

Rogers considers six functions of fragmentation which are related to psychological and literary aspects, an appeal to the reader's own psychology, the stimulation of defensive adaptations, representation and defence, and the institution of aesthetic distance (Rogers 172). In these narratives, it will be shown that in the entire ST *gesamtkunstwerk*, decomposition is only used to expose and investigate various psychological facets of the human condition.

Shadows that must be physically reintegrated
Two Kirks
In "The Enemy Within" (Penn), Kirk is accidentally doubled by a transporter accident (Grech, "The Trick"), "his negative side, which you call hostility,

lust, violence, and his positive side, which Earth people express as compassion, love, tenderness." Woods and Harmon explain that the former manifestation is "created by the malfunction of a product of advanced technology, (…) a duplicate characterised by violence and anger" (172), while they incorrectly equate the latter with the everyday Kirk persona. Due to the division, the positive half finds himself

> rapidly losing the power of decision (…) what is it that makes one man an exceptional leader? We see indications that it's his negative side which makes him strong, that his evil side, if you will, properly controlled and disciplined, is vital to his strength. Your negative side removed from you, the power of command begins to elude you. (…) If your power of command continues to weaken, you'll soon be unable to function as Captain. You must be prepared for that (Penn).

The negative half initially poses as the real captain, adopting a changeling role, and is quickly unmasked. The positive half has insight: "My negative self is under restraint in Sickbay. My own indecisiveness growing. My force of will steadily weakening." He also realises that reintegration is crucial: "I have to take him back inside myself. I can't survive without him. I don't want him back. He's like an animal, a thoughtless, brutal animal, and yet it's me. Me."

Such scenes permit dramatic clarification such that "when an author wishes to depict mental conflict within a single mind a most natural way for him to dramatize it is to represent that mind by two or more characters" (Rogers 29). The ship's doctor reassures the captain

> We all have our darker side. We need it! It's half of what we are. It's not really ugly, it's human. (…). A lot of what he is makes you the man you are. (…) Without the negative side, you wouldn't be the Captain. You couldn't be, and you know it. Your strength of command lies mostly in him. (…) You have the goodness. (…) The intelligence, the logic. It appears your half has most of that, and perhaps that's where man's essential courage comes from. For you see, he was afraid and you weren't.

This is confirmed by the negative half who pleads against reintegration: "Please, I don't want to. Don't make me. Don't make me. I don't want to go back. Please! I want to live!" The positive half attempts to reassure the negative half: "You can't hurt me. You can't kill me. You can't. Don't you understand? I'm part of you. You need me. I need you."

After reintegration through the reuse of the transporter, the captain muses: "I've seen a part of myself no man should ever see (…) The impostor's back where he belongs. Let's forget him."

Incidentally, Kirk is also physically doubled for nefarious purposes by shape-shifting aliens on two occasions (Wallerstein, "Whom Gods Destroys;"Nimoy, "Star Trek IV: The Voyage Home").

The Human and Klingon halves of B'Elana Torres

In another episode, a half-Klingon and half-human female engineer (B'Elana Torres) is split into her two halves, one completely human and one typically Klingon, aggressive and warrior-spirited, a proxy for B'Elana's shadow. The human half confesses "when they extracted my Klingon DNA, they turned me into some kind of a coward," alluding to the unfortunate consequences of a splitting off of a vital archetype, and yet, when she confronts the Klingon half, she accuses her:

> That's the way you respond every situation, isn't it? If it doesn't work, hit it. If it's in your way, knock it down. No wonder I got kicked out of the Academy. (…) Your temper has gotten me into trouble more times that I can. Listen to me. Listen to us. This is ridiculous. Do you realise we're each fighting with our

self. (…) Out of control. Just leaping into action before you think things through (Kolbe, "Faces").

This supports Rogers' assertion that the "double represents both qualities he hates in himself and attributes he lacks and desires to have" (Rogers 17). B'Elana' two halves are eventually integrated, just as Captain Kirk was.

Drug and device induced shadows

Drugs may provoke the appearance of repressed shadows, and the Cardassian tailor/spy/assassin Garak experiences a psychotic break after accidentally ingesting a "psychotropic compound that was affecting his nervous system." His colleagues remark: "He looks so peaceful. It's hard to believe he's the same man who attacked us." The doctor explains that "[i]n a way, he's not. The drug brought out the worst parts of him and allowed them to take over. He wasn't in control of his actions" (Vejar, "Empok Nor").

Similarly, the sword of the revered legendary Klingon warrior Kahless brings out the very worst in two Klingons who discover it, leaving the Trill companion untouched. The Klingons display aggression, violence, hatred, jealousy, delusions of grandeur and outright paranoia. They realise that they are not"ready for it. The Sword turned you and me against each other. Imagine how it would divide the Empire. […] The Sword is not meant for us. It was never our destiny to find it. […] When it is destined to be found, it will be" (Burton, "The Sword of Kahless").

Shadows that cannot be physically reintegrated
Physical integration may not always be required or necessary but may occur psychologically if the shadow is extracted and examined and lessons thereby learnt.

Two Spocks
The very first Trek novel "Spock Must Die!" was written by noted author James Blish, who was responsible for the ST Original Series novelisations. In this story, a duplicate Spock is created and transported, but the transporter signal bounces off a thought-reflecting field and creates a mirror Spock, and after the eventual death of the latter, the original Spock confirms that

> his motives, his morals, his loyalties were all the opposite of my own. (…) he had at his command (…) all my intimate knowledge of the Enterprise, its crew, and the total situation. And hence, I knew that he was a terrible danger to us all, and under no circumstances could he be negotiated with.

He had to be eliminated, (…) there was no other possible course.

Spock and his half-brother

Sybok, Spock's half brother, is fully Vulcan. He was exceptionally gifted, possessing great intelligence. It was assumed that one day he would take his place amongst the great scholars of Vulcan. But he was a revolutionary. […]The knowledge and experience he sought were forbidden by Vulcan belief. […]He rejected his logical upbringing. He embraced the animal passions of our ancestors.[…]He believed that the key to self-knowledge was emotion, not logic. […] When he encouraged others to follow him, he was banished from Vulcan, never to return.

The subversive Sybok may therefore be viewed as Spock's (and the entire Vulcan race's) shadow (Shatner, "Star Trek V: The Final Frontier"). Moreover, Sybok steals the *Enterprise* in order to journey to the centre of the galaxy, where he believes that God resides. Instead, he finds a shadow, an evil and cruel superbeing who is an impostor. Sybok sacrifices his life in a struggle with this being and both are killed.

Two Kirks and two Picards

An android copy of Kirk is created who must be killed so as to avoid spreading other androids throughout the galaxy (Goldstone, "What Are Little Girls Made Of?"). Similarly, a doppelganger Picard takes over the *Enterprise* and orders dangerous manoeuvres until removed by his creators (Kolbe, "Allegiance").

Two Datas

Data is a sentient android, an artificial life-form. His creator had first assembled another android called Lore. Since this android was stronger and had a better brain than humanity, Lore became emotionally unstable and malevolent toward his human creators, in all respects the opposite of Data, who wishes to become more human (Grech, "Pinocchio"). Lore attempts to subvert Data and destroy the *Enterprise* (Bowman, "Datalore," Bole, "Silicon Avatar"), kills his creator (Bowman, "Brothers") and eventually takes over a group of Borg, cybernetic organisms. He unethically and callously experiments on them, replacing parts of their organic brains with artificial positronic components (Singer, "Descent"), and is eventually dismantled after being defeated.

Two Datas with two Picards

In *Nemesis* (Baird), the new leader of the Romulan Empire is a chronologically younger clone of Picard. This clone plans to destroy Earth:

If you had lived my life and experienced the suffering of my people, you'd be standing where I am (…) I can't ...fight what I am! (…) I'll show you my true nature. Our nature. And as Earth dies, remember that I will always, forever, be Shinzon of Remus! And my voice shall echo through time long after yours has faded to the dim memory.

Picard remonstrates, with existential angst, confirming that "[w]hen an author portrays a protagonist as seeing his double, it is […] a result of his sense of the division to which the human mind in conflict with itself is susceptible" (Rogers 29):

Look at me, Shinzon. ...Your heart, your hands, your eyes are the same as mine. The blood pumping within you, the raw material is the same. We have the same potential. (…) It can be the future. Buried deep within you, beneath all the years of pain and anger there is there is something that has never been nurtured. The potential to make yourself a better man, and that is what it is to be human. To make yourself more than you are. ...Oh yes, I know you. ...There was a time you looked at the stars and dreamed of what might be.(…). I see what you could be. ...The man who is Shinzon of Remus and Jean Luc

> Picard could never exterminate the population of an entire planet! He is better than that! (…) what will he do with that life? Waste it in a blaze of hatred? There is a better way. (…) You still have a choice! Make the right one now!

Since the clone elects to attempt to destroy Earth, he is killed by Data.

Moreover, earlier in the same film, an android that had also been fashioned by Data's creator is discovered, an android called B-4 that had been suborned by Shinzon, possibly because "he's a prototype, a lot less sophisticated than" Data. However, Data interfaces with B-4 and downloads his memories into B-4, in the hope that if Data's

> memory engrams are successfully integrated into his positronic matrix, he should have all my abilities (…) with my memory engrams he will be able to function as a more complete individual. (…) I believe he should have the opportunity to explore his potential.

In this way, by merging with and counter-suborning his evil alter-ego, Data not only neutralises the threat presented by B-4, but also lives on within B-4 when he is eventually killed while saving his Captain and the*Enterprise*.

Real and virtual Datas

Computer software may also directly manifest as the shadow. While experimenting with the possibility of utilising Data's brain as an emergency backup for the ship's computer, Data inadvertently overwrites and corrupts software and systems within the ship's computer, dangerously disrupting a holodeck cowboy simulation of the far west. In this simulation, Data fragments into multiple villains, retaining the real Data's speed and strength, formidable and seemingly invincible opponents. Interestingly, in the real world, Data's software also becomes corrupted with cowboy slang and mannerisms. The holodeckDatas are defeated before they can seriously injure the individuals within the holodeck, and the situation is ultimately resolved when both the ship's computer and Data are purged of the corrupted software (Stewart, "A Fistful of Datas").

Two O'Briens

The chief engineer is cloned and infused with the original's memories. However, while he has no inkling that he is not the original, he is biologically programmed to be triggered to sabotage a crucial set of peace talks. This "manifest double" is physically real and even on medical testing indistinguishable from the original, with "the projected self being not merely a similar self but an exact duplicate" (Rogers 19). He is killed during his efforts to appraise his superiors that something is amiss on Deep Space Nine (Landau, "Whispers").

A murderer in one's past

Deep SpaceNine's science officer, JadziaDax is a "Trill," a humanoid species that physically host "symbionts," long-lived, slug-like creatures which mentally integrate with the host survive its death to be re-implanted into another Trill, in effect, "serial binaries" (Lundeen and Wagner 74).

Jadzia'ssymbiont was once hosted by a murder, whose memories have been repressed, just like a repressed shadow. This almost causes Jadzia's death, which is forestalled when the memories of the murderer are accepted and re-integrated (Bole, "Equilibrium"). In a later episode, Jadzia furthers her process of individuation by undergoing a Trill ritual that permits her to meet the symbiont's previous hosts through a temporary process of memory projection of each individual dead host's memories into the mind of her closest friends, a deliberate decomposition of her memories.

Decomposition thus "involves the splitting up of a recognizable, unified psychological entity into separate, complementary, distinguishable parts represented by seemingly autonomous characters" (Rogers 5). Afterward, Jadzia mused: "[i]t forced me to deal with some things about myself I've never really faced" (Bole, "Facets").

A Docile Dax

The crew of Deep Space Nine discover that their wishes can become manifest as tangible persons and the resident medical Dr. Julian Bashir inadvertently doubles the beautiful but unattainable Jadzia, creating a fantasy woman who "really is submissive." The original sardonically asks the doctor "[i]s that how you want me to be, Julian? So submissive?" (Legato, "If Wishes Were Horses"). The double eventually disappears.

Dr. Bashir's dreaming archetypes

After cranial trauma, Bashir experiences a hallucination wherein his archetypes decompose, manifesting as the rest of the senior crew who "embody different aspects of my personality, different voices inside my head." These individuals separately represent "doubt and [..] disbelief [...] aggression [...] sense of suspicion and fear [...] confidence and sense of adventure [...] professionalism and [...] skill" (Singer, "Distant Voices"). It is this episode that most classically results in "the splitting up of a recognizable, unified psychological entity into separate, complementary, distinguishable parts represented by seemingly autonomous characters" (Rogers 10).

Mental projections

A serious and introverted young woman who is unhappy in her marriage creates a physical "psychoprojective" copy of herself who is her opposite in character (Singer, "Second Sight"). This double eventually disappears.

The Emergency Medical Hologram
Accidents that unintentionally elicit the shadow can also happen to sentient but nonhuman beings. The Emergency Medical Hologram on. *Voyager* interviews historical re-creations as part of his

> personality improvement project. I've been interviewing the historical personality files in our database. Socrates, da Vinci, Lord Byron, T'Pau of Vulcan, Madame Curie, dozen of the greats. Then I select the character elements I find admirable and merge them into my own programme. (…) An improved bedside manner, a fresh perspective on diagnoses, more patience with my patients (Singer, "Darkling").

But this creates problems:
> Lord Byron (…) A creative, poetic genius. (…) But Byron was also emotionally intense, even unstable. (…) And T'Pau. She was a diplomat, a judge, a philosopher. (…) And utterly ruthless in her application of that logic. (…) A lot of the historical characters you chose have this dark thread running through their personalities.

The resultant creation exhibits the combined shadow of all of the integrated personalities, in effect "a new personality, from the subroutines." The new and malevolent doctor explains:

> I was born of the hidden, the suppressed. I am the dark threads from many personalities. (…) None of whom could face the darkness inside so they denied me, suppressed me, frightened of the truth. (…) That darkness is more fundamental than light. Cruelty before kindness. Evil more primary than good. More deserving of existence.

He comments on his previous existence serving as the ship's doctor:

> A little bit of this, a little bit of that. (…) What a hollow excuse for a life. Servile, pathetic, at the beck and call of any idiot who invokes his name. The thought of him sickens me. (…) he repulses me. (…) Because he's as weak as the rest of you. He fails to understand the power of his own holographic nature. He is detestable. There's not enough room inside for both of us. One must die. I deserve to exist more than your Doctor does.

He boasts of his attempts at murder, treachery and outright betrayal in Nietzschean fashion:

I am beyond considerations of wrong and right. Behavioural categories are for the weak, for those of you without the will to define your existence, to do what they must, no matter who might get harmed along the way. (…) I fear nothing, no-one.

But when he attempts to completely expunge the old personality, he finds that he cannot: "d]elete the Doctor and you go too. The subroutines are all inter-connected," reinforcing the notion of the necessity of all archetypes for the existence of the whole.

When the extra personality subroutines are finally eliminated, the composite shadow is expunged, and like Kirk, the doctor comments "good riddance." The overall episode is a clear reference to Stevenson's *The Strange Case of Dr. Jekyll and Mr. Hyde*, albeit with a happy ending.

Stalemate

The *Enterprise* crew discover two almost identical aliens, one of who is sane and comes from an alternate antimatter universe and one from our universe and who is insane. The normal universe alien had

lost his mind. When our people found a way to slip through the warp and prove another universe, an identical one, existed, it was too much for him. He could not live knowing that

I lived. He became obsessed with the idea of destroying me.

The story indicates that should the two meet outside a corridor that links the two universes, then both universes would be destroyed, and "[t]he fact that it meant his own destruction, and everything else, meant nothing to him." The rational alien has Kirk permanently trap both aliens in the corridor, a sacrifice leading to a permanent struggle. "Is it such a large price to pay for the safety of two universes?" (Oswald, "The Alternative Factor").

The Shadow unleashed by excessive powers

In various ST episodes, beings with disproportionately large powers or who suddenly acquire new powers may oppress others who are weaker, or may find themselves overwhelmed by their new-found powers and make incorrect decisions.

Children are immature and egocentric beings whose shadow is more easily brought to the fore if they are undisciplined. Charlie (Dobkin, "Charlie X"), Trelane (McDougall, "The Squire of Gothos"), Amanda Rogers (Scheerer, "True Q") and the surviving children of a scientific colony (Chomsky, "And the Children Shall Lead") all have extraordinary powers with which they threaten the crew of the *Enterprise*. Moreover, the latter group have been suborned by an "evil [that] is awaiting a catalyst to set it again into motion and send it marauding across the galaxy." These children have been given the ability to summon "the enemy from within," paralysing all of the adults in the crew at will by recalling their deepest subconscious fears. Kirk confronts the evil after his "beast is gone. It lost its power in the light of reality." Evil responds:

> My followers are strong and faithful and obedient. That's why we take what is ours wherever we go. (...) You will be destroyed. I would ask you to join me, but you are gentle, and that is a grave weakness. (...) your strength is cancelled by your gentleness. You are full of goodness. (...) You must be eliminated. (...) Together we can raise armies of followers. (...) We shall exterminate all who oppose us! Our purity of purpose cannot be contaminated by those who disagree, who will not co-operate, who do not understand. They must be annihilated. (...) You will be swept aside to make way for the strong.

But the children reject the evil which then obligingly vanishes from the *Enterprise*.

Adult superhuman individuals may also whimsically indulge their shadow, and the quasi-supernatural being Q is a typical example, Loki-like playing tricks on the *Enterprise* crew (Corey, "Encounter at Farpoint").

Human, Klingon, Vulcan, and Romulan collective shadows

The shadow within the collective unconscious has also manifested in the canon, either irrupting after long periods of dormancy through suppression, or as a way of life for an entire race. This is often manifest in a race's myths, since "[e]very culture has its demons. They embody the darkest emotions of its people. Giving them physical form in heroic literature is a way of exploring those feelings" (Landau, "Heroes and Demons")

Human

The *Enterprise* finds a colony governed by a computer (Landru) where everyone is at peace, and violence and all other negative emotion are suppressed. To Landru"[t]he good is the harmonious continuation of the Body. The good is peace, tranquillity," with an almost bovine, excessively calm existence. However, Landru allows the populace free rein for their emotions in a recurrent "Festival," a Bacchanalian orgy of rape, fighting, destruction and looting. Kirk rejects this *status quo*, accusing Landru:

> The body is dying. You are destroying it. What have you done to do justice to the full

> potential of every individual of the Body?
> Without freedom of choice, there is no
> creativity. Without creativity, there is no life.
> The body dies.

This reiterates the notion that the shadow is
necessary for creativity and individuality, and
Landru accepts the accusations and obligingly
destroys itself, freeing the colony of its rule (Pevney,
"The Return of the Archons").

Moreover, Starfleet itself contains a shadow called
Section 31 headed by the enigmatic Mr. Sloan. This
is a nefarious organisation that has an enigmatic
relationship with Starfleet and the Federation. "The
Federation claims to abhor Section Thirty One's
tactics, but when they need the dirty work done, they
look the other way. It's a tidy little arrangement,
wouldn't you say?" (Brooks, "The Dogs Of War").
Naturally, this arrangement is repugnant to the
protagonists who find out that Section 31, went so far
as to plan genocide: "[t]his organisation, this thing
that's slithered its way into the heart of the
Federation, has to be destroyed" (Posey, "Extreme
Measures").

Interestingly, Sloan commits suicide when forced to
divulge information related to his genocidal plans,
and on using a medical device to enter his mind, two
protagonists find themselves in a simulated physical
environment. When helped by a Sloan who seems to

have repented and is intent on helping them, another, black-clad Sloan manifests and kills the helpful Sloan, thereby functioning as a shadow that displaces the original (Posey, "Extreme Measures").

The Borg

The Borg, a race of cybernetic humanoid organisms that comprise a hive-mind collective, are the very antithesis of humanity and the United Federation of Planets (Corey, "Encounter at Farpoint"). They are represented as ugly mechanised Marcusian zombies, inhabiting places that are dark and dank, lurching into movement in order to assimilate independent individuals and even entire species into the collective, with loss of individuality, will and personal control. The Borg may therefore be viewed as the "shadow of the Federation" (Lundeen and Wagner 206).

Klingon and human

A mysterious energy being enters the *Enterprise* and somehow precipitates a "magnification of the basic hostilities between humans and Klingons," unleashing aggressive shadows that provoke violent outbreaks between the ship's crew and a group of Klingons who have been recently captured. The crews realise that "it is by design that we fight. We seem to be pawns." This is because the being

> subsists on the emotions of others. (…) strengthened by mental irradiations of

hostility, violent intentions. (…) It exists on the hate of others. (…) [I]t has acted as a catalyst, creating this situation in order to satisfy that need. It has brought together opposing forces, provided crude instruments in an effort to promote the most violent mode of conflict. (…) And kept numbers and resources balanced, so that it can maintain a constant state of violence.

Moreover, the being repairs violently wounded individuals and brings the dead back to life,

[i]n the heart. In the head. I won't stay dead. Next time I'll do the same to you. I'll kill you. And it goes on, the good old game of war, pawn against pawn! Stopping the bad guys. While somewhere, something sits back and laughs and starts it all over again.

Everyone eventually comes to the realisation that

all hostile attitudes on board must be eliminated. The fighting must end and soon. (…) Or we're a doomed ship, travelling forever between galaxies, filled with eternal bloodlust, eternal warfare. (…) For the rest of our lives. A thousand lifetimes. Senseless violence, fighting, while an alien has total control over us.

When all calm down and cease fighting, the being considerately leaves the ship and the shadows are controlled (Chomsky, "Day of the Dove").

Vulcan and Romulan

The Vulcan shadow is most obviously precipitated by the *ponfarr* mating ritual which occurs every seven years (Pevney, "Amok Time") and has been reviewed in depth (Grech, "Irruption"). This constitutes an extreme physiological storm that rages through the Vulcan body in an uncontrollable and inexorable manner, such that the shadow is violently unleashed. "Perhaps it's the price they pay for having no emotions the rest of the time."

Since Vulcans are calm individuals who strive to maintain emotional equanimity, the Romulan race may be considered a reification of the Vulcan shadow. This is because Romulans were originally Vulcans who refused to accept Surak's philosophy of logic to suppress emotion, and are colloquially known as "those who marched beneath the Raptor's wings" (Grossman, "The Forge"). Having abandoned Vulcan, they colonised the planet Romulus and conquered neighbouring races, developing in semiotic counterpoint to their Vulcan cousins. Romulans are passionate, cunning and opportunistic and may be perceived by the naïve as being

> very moral (…). They have an absolute
> certainty about what is right and what is
> wrong, who is a friend and who is an enemy,
> a strict moral compass which provides them
> with a clarity of purpose. (…) their sense of
> purpose, their passion and commitment, (…)
> very compelling (Beaumont, "Face of the
> Enemy").

Romulans may be viewed as shadow manifestations in two ways. Firstly, the entire Romulan empire and ethos may be viewed as the Vulcan shadow, a Levi-Straussian antithesis of the moral and dispassionate Vulcans who value "integrity and personal honour" (Lucas, "The Enterprise Incident"). Or they maybe regarded as Vulcans who have acceded to their shadow and allowed it free reign in controlling their destiny.

Moreover, Romulans are also very different from Vulcans in their mating habits and a female Romulanstarship commander rubs Spock's nose in this while attempting to corrupt him, insinuating that sex does not necessarily have to bow to a seven year mating cycle thereby appealing to his human half.

> We have other inducements. (...) Romulan
> women are not like Vulcan females. We are
> not dedicated to pure logic and the sterility of
> non-emotion. Our people are warriors. Often

savage. But we are also many other pleasant
things. (…) As a Vulcan, you would study it.
As a human you would find ways to
appreciate it (Lucas, "The Enterprise
Incident").

Thus, in turn, over the history of the ST timeline,
both Klingons and Romulans participate as
collective shadows (Tyrrell).

The Mirror Universe

A parallel or alternate universe is a hypothetical
universe that is entirely separate in space and time
from our own. The existence of such universes is
predicted by Everett's "many-worlds" interpretation
of quantum mechanics, the multiverse. There are a
potentially infinite number of parallel universes in
the ST canon (Wiemer, "Parallels") but the *Mirror
Universe* is most commonly interacted with. The
Mirror Universe is a sinister locus where events
happen as if seen "through the looking glass"
(Livingston, "Crossover"). This universe has been
invoked in one episode of *Star Trek: The Original
Series*, five episodes of *Star Trek: Deep Space Nine*,
and two episodes of *Star Trek: Enterprise*.

The *Mirror Universe* arises due to the formation of a
totalitarian Terran Empire instead of the peaceful
United Federation of Planets, an empire in which
dominant humanity enslaves alien worlds and

warfare and rebellion are continual, with humanity eventually overthrown by another empire. Characters in the *Mirror Universe* correspond to those in this universe, but are usually more aggressive, distrustful, treacherous and opportunistic. Such characters serve to reify shadows of their canonical universe counterparts and provide many opportunities for contrasting the optimistic beneficence of the United Federation of Planets with the depressing and cheerless outlook of the Mirror Universe. Kirk urges the former values upon mirror Spock:

> The illogic of waste, Mister Spock. The waste of lives, potential, resources, time. I submit to you that your Empire is illogical because it cannot endure. I submit that you are illogical to be a willing part of it. (...) If change is inevitable, predictable, beneficial, doesn't logic demand that you be a part of it? (...) one man can change the present. (...) What will it be? Past or future? Tyranny or freedom? It's up to you. (Daniels, "Mirror, Mirror").

In the mirror universe, a human slave wonders about a human doctor who has entered the mirror universe from this one:

> This man is a doctor where he comes from. And there's an O'Brien there, just like me...

Except he's some kind of high-up chief of operations. And they're Terrans. Can you believe that? Maybe it's a fairy tale he made up, but it started me thinking how each of us might have turned out, had history been just a little different" (Livingston, "Crossover")

Actual Subtraction of Evil
Mental Displacement
An interesting episode postulates an individual who can coerce another into becoming a receptacle for his shadow, an "ability to channel (…) darker thoughts, (…) unwanted emotions, to others, leaving me unencumbered." The recipient becomes angry and aggressive, hypersexual, ill and ages rapidly and dies, while the sender, a treaty negotiator, becomes "focused, centred, free of disquieting thoughts." Picard naturally rails against this

> wantonly immoral act because you think it is connected to some higher purpose. That does not justify brutalising her, nor any of the others you have used (…)You're a coward (…). You exploit the innocent because you're unwilling to shoulder the burdens of unpleasant emotions.

The situation is resolved when a means is found to have "the flow of negative emotions and feelings

(…) reversed back to him," a fatal outcome (Kolbe, "Man of the People").

Physical Removal

An entire race somehow manages to expel its collective shadow, with tragic consequences for the *Enterprise*. This race "perfected a means of bringing to the surface all that was evil and negative within. Erupting, spreading, connecting. In time it formed second skin, dank and vile (…) I do not serve things evil. I am evil" (Scanlan, "Skin of Evil").

This being was discarded on an empty planet and left behind for the unwary to stumble upon. In doing so, the shadow reminisces that his derivative race became "creatures whose beauty now dazzles all who see them. They would not exist without me," such that "evil is accentuated and isolated in a diabolical other self" (Rogers 33).

He further elaborates: "I am a skin of evil left here by a race of Titans who believed if they rid themselves of me, they would free the bonds of destructiveness" (Scanlan, "Skin of Evil"). The *Enterprise's* Chief Security Officer is killed, "an empty death. A death without purpose (…) a senseless death" (Carson, "Yesterday's Enterprise") by "a malevolent entity (…) killed as a demonstration of the creature's power, without provocation" (Scheerer, "Legacy").

However, the creature has been abandoned "since they left. A very long time. (…) A long time to be alone," intimating a lonely existence (Scanlan, "Skin of Evil"). Picard defies it:

> Where do you want to go? Do you want to try to find those who left you here? (…) Because you feel unworthy. (…) A great poet once said, 'all spirits are enslaved that serve things evil.' (…) So here you are. Feeding on your own loneliness. Consumed by your own pain. Believing your own lies. You say you are true evil? Shall I tell you what true evil is? It is to submit to you. It is when we surrender our freedom, our dignity, instead of defying you. (…)But you will still be here. In this place. For ever. Alone. Immortal. (…) That's your real fear. Never to die. Never again to be united with those who left you here. (…) I'm not taking you anywhere.

And to forestall any other being becoming ensnared by this physical incarnation of evil, the planet is "declared off limits."

Discussion

Rogers commences his study with the hypothesis that humans are inherently double or multiple in their nature. Furthermore, Rogers concludes that authors inadvertently divulge their own intrinsic repressed selves when creating narratives that expose hidden archetypes. Rogers moreover observed that "the value of discussing multiple personality in connection with multiple decomposition is largely analogical. There do not seem to be many literary works which exhibit a precise correspondence to the phenomenon of true dissociation" (93). However, in contrast with mainstream literature, in SF narratives such as those outlined above, "dual and multiple fragmentation" (5) are relatively easily accomplished, through technological or other novums.

Doubling and fragmentation in ST has been shown to result in various manifestations of the shadow, which seems "to contradict the notion of a stable, bounded self marked by continuity over time (Wagner and Lundeen 79)." However, doubles and greater multiples may arise from "a bifurcation of space/time that has little to do with moral duality" (Lundeen and Wagner 72), such as Picard meeting himself in a wrinkle in time (Scanlan, "Time Squared") and the *Voyager* crew meeting other versions of themselves from an alternate space-time continuum (Livingston, "Deadlock"). Such "eigenshadows" (Rogers et al) manifest vectors that parallel the somehow privileged original and have no ethical or moral superiority or inferiority.

In all of the other narratives outlined above, there is no "doubling by multiplication," a fragmentation resulting in the creation of multiple characters that represent a single concept (Rogers 4), presumably because this would not add significantly to the dramatic tensions in ST episodes, and would therefore be superfluous. Therefore doubling and fragmentation manifest as shadows that are largely opposed to the individual Starfleet officer, creating an "opposing self" (Rogers 62) that is in conflict with the overall precepts of the United Federation. These include Aristotlean moral virtue ethics reinforced by Kantian deontological principles adherent to the categorical imperative, such that actions are carried out that are good in themselves and therefore morally worthy, irrespective of the eventual consequences (Barad and Robertson). This reaffirms Rogers' view that "[t]he conventional double is [...] some sort of antithetical self, usually a guardian angel or tempting devil. Critics oriented toward psychology view the diabolic double, which predominates, as a character representing unconscious, instinctual drives" (Rogers 2), that is, the shadow.

Federation values therefore inescapably engender Kierkegaardian existentialism, the realisation that the responsibility for giving meaning to one's life is belongs solely to the individual, modified by the Sartrean viewpoint that there can be no deferral of responsibility to divine beings, to others or even to human nature and that one is solely responsible for one's actions (Eberl and Decker).

Aristotlean friendship of goodness is also fostered with a deep and abiding friendship and loyalty. Moreover, an epicurean lifestyle is pursued, when temperance and prudence used as a guide to decide which pleasures to seek and which to avoid.

The shadow thus manifests as the evil Manichean counterpart to the principles of the Federation, symmetrically completing a Lévi-Straussian dyadism, a convenient construct since "[e]veryone carries a shadow, and the less it is embodied in the individual's conscious life, the blacker and denser it is" (Jung, "Psychology and Religion" 131). Indeed, Manichaeism is ubiquitous in ST with episodes often playing themselves out as subtle (and sometimes not so subtle) morality plays, and in the abovementioned narratives, the shadow acts out the role of the evil complement.

It is thus that *Star Trek*'s popularity "can (…) more easily be understood by referring to basic and universal psychic structures" (Blair 311) as physical beings who can be readily comprehended and appreciated by viewers.

The shadow in ST also replaces the villains in traditional fairy tales, time-honoured monsters that have been banished by science and technology, so that speculative journeys to distant planets have led authors to populate "these new unknowns with monsters and ogres that could well be the close relatives of the trolls and ogres of folklore fame. In that sense […] SF is modern folklore" (Schelde 4), to the extent that SF has been cogently argued as acquiring the role of modern myth (Kappell). Thus, "just as the principal character projects his malevolent impulses onto his double, thereby disclaiming any responsibility for such impulses, so is the reader easily able to shunt off the guilt he unconsciously shares with the evil protagonist" (Rogers 33), an identification and projection with which fans readily associate. In this way, the shadow in SF becomes the modern myth's equivalent of "spells, demonic possession, soul loss anddoubles" (Lundeen and Wagner 74).

The shadow, this "double is an evil version of the self, the repository of all the personal traits that one ordinarily refuses to confront and may actively deny, but which remain incorrigibly present in the recesses of the personality" (Lundeen and Wagner 70), with potentially devastating consequences. The shadow serves several purposes in the canon, and all were neatly summarised by Jung himself. The shadow reminds us that man is, on the whole, less good than he imagines himself or wants to be. Everyone carries a shadow, and the less it is embodied in the individual's conscious life, the blacker and denser it is. (…) if it is repressed and isolated from consciousness, it never gets corrected (Jung, "Psychology and Religion" 131).

We are also reminded that this is a powerful archetype, a "shadow side (…), consisting not just of little weaknesses and foibles, but of a positively demonic dynamism. (…) a raging monster (Jung, "On the Psychology of the Unconscious" 35). Moreover, there is a deliberate lack of insight in that "[h]aving a dark suspicion of these grim possibilities, man turns a blind eye to the shadow-side of human nature. (…), he even hesitates to admit the conflict of which he is so painfully aware" (ibid.), a deliberate rejection of the possibility of the very existence of the shadow.

This is because our mind retains ancient primordial vestiges, such that "[w]e carry our past with us, to wit, the primitive and inferior man with his desires and emotions, and it is only with an enormous effort that we can detach ourselves from this burden" (Jung, ""Answer to Job" 12), a remnant of our precedents of which we are simultaneously embarrassed and ashamed.

Indeed, one must deliberately seek out one's shadow since "no one can become conscious of the shadow without considerable moral effort. To become conscious of it involves recognizing the dark aspects of the personality as present and real. This act is the essential condition for any kind of self-knowledge" (Jung, "Aion" 14).

This is directly alluded to in ST, when the *Enterprise*'s counsellor (psychologist) notes that "[s]ometimes it's healthy to explore the darker sides of the psyche. Jung called it owning your own shadow" (Conway, "Frame of Mind").

The search for and reconciliation with the shadow must be an active process. However, with the best will in the world, "the acceptance of the shadow-side of human nature verges on the impossible. Consider for a moment what it means to grant the right of existence to what is unreasonable, senseless, and evil!" (Jung, "Psychotherapist or the Clergy" 528). Generally speaking, the individual's "[f]ailure to consciously acknowledge our own material and to unconsciously project it onto others, is to contribute to the sum total of evil in the world" (Woods 171), which imposes an actual responsibility for the individual to reconcile the self with the shadow.

However, the shadow may also be unintentionally overlooked by individuals who "have no suspicion whatever of the internecine war raging in their unconscious. (…) people who are utterly unaware of their actual conflicts" (Jung, "New Paths in Psychology" 425).

A deliberate or inadvertent rejection of the shadow results in a refutation of individuation, the process whereby unconscious archetypes are merged and integrated in order to produce a stable individual with a balanced personality, with possible destabilisation of the self. Man "constantly needs the renewal that begins with a descent into his own darkness" (Jung, "Mysterium" 334). Rebuffing the shadow carries significant risks, even to humans, as noted by aliens, since as noted by the Cardassian Garak, "under your Federation mask of decency and benevolence, you're a predator" (Vejar, "Empok Nor"). Moreover, a Ferengi bartender notes

I still don't want you anywhere near them. Let me tell you something about humans, nephew. They're a wonderful, friendly people as long as their bellies are full and their holosuites are working. But take away their creature comforts, deprive them of food, sleep, sonic showers, put their lives in jeopardy over an extended period of time, and those same friendly, intelligent, wonderful people will become as nasty and as violent as the most bloodthirsty Klingon. You don't believe me? Look at those faces. Look in their eyes. You know I'm right, don't you? (Kolbe, "Siege of AR-558"). This was averred by Jung "a small evil becomes a big one through being disregarded and repressed" ("A Psychological Approach to the Dogma of the Trinity" 286). Prolonged neglect may have catastrophic consequences.

> The change of character (…) is amazing. A gentle and reasonable being can be transformed into a maniac or a savage beast. (…) we are constantly living on the edge of a volcano, and there is, so far as we know, no way of protecting ourselves from a possible outburst that will destroy everybody within reach (Jung, "Psychology and Religion" 25).

This is also depicted in ST, and in an episode when the crew forget who they are while retaining their practical skills, it is observed that "[w]hen you have no memory of who you are, or who anybody else is, […] we might do the things that we've always wanted to do" (Landau, "Conundrum"). Moreover, several of these episodes affirm Jung's contention that "in spite of its function as a reservoir for human darkness—or perhaps because of this—the shadow is the seat of creativity" such that "the dark side of his being, his sinister shadow ... represents the true spirit of life as against the arid scholar," (Jung, Memories 262) eliciting yet another crucial need for individuation. Moreover, individuation requires that all memories and experiences, agreeable or otherwise, are accepted, since "if you want to know who you are, it's important to know who you've been" (Bole, "Equilibrium").

These narratives also demonstrate that when "when a man has found peace with himself and the world it is indeed a noteworthy event" (Jung, "Analytical Psychology and Weltanschauung" 693), one to be actively sought.

Naturally, members of the United Federation of Planets and of Starfleet function as heroes, with shadows defeated by integration or destruction. Jung views this as "the hero's main feat [...] to overcome the monster of darkness: it is the long-hoped-for and expected triumph of consciousness over the unconscious" (Jung, "The Psychology of the Child Archetype" 284).

Furthermore, a hopeful theme constantly resurfaces in ST, the anticipation that an accommodation of some kind will be reached with our shadow, both individually and racially. At the individual level, a protagonist is exhorted: "[d]on't deny the violence inside of you, Kira. Only when you accept it can you move beyond it"

(Lynch, "Battle Lines"). And an acknowledgment of the wider, racial potential for the emergence of the shadow is postulated by Picard:

> Earth was once a violent planet, too. At times, the chaos threatened the very fabric of life, but (…) we evolved. We found to find better ways to handle our conflicts. But I think no one can deny that the seed of violence remains within each of us. We must recognise that, because that violence is capable of consuming each of us. (Wiemer, "Violations").

This optimism in ST also extends to emergent phenomena that arise from the *Enterprise*'s collective unconsciousness. When an intelligence is born out of the ship's computer, Picard observes that

> [t]he intelligence that was formed on the Enterprise didn't just come out of the ship's systems. It came from us. From our mission records, personal logs, holodeck programs, our fantasies. Now, if our experiences with the Enterprise have been honourable, can't we trust that the sum of those experiences will be the same?(Bole, "Emergence").

In conclusion, these narratives establish clear rules: that the shadow is vanquished by individuation or by being banished or destroyed or isolated after one comes to term with it, such that "[t]he traditional binary opposition between 'Us' and 'Them' becomes what Derrida calls a 'Crisis of versus'(…) 'They're us…we're them.'" (Badmington 32).

These episodes also function as cautionary tales with a dual reminder that "[h]e who fights with monsters should be careful lest he thereby become a monster. And if thou gaze long into an abyss, the abyss will also gaze into thee" (Nietzsche 63). However, these episodes also evoke the optimistic expectations that in the not too distant future, we will become reconciled with our shadow, a complete process of individuation that will allow us to become our true (and therefore better and happier) selves (Winnicott). Acknowledgments

I would to acknowledge the invaluable transcripts at Christina Lucking's site, the "one-stop site for who said what and when on Star Trek" (http://www.chakoteya.net), as well as Memory Alpha, the collaborative wiki, "encyclopedia and reference for everything related to Star Trek" (http://www.chakoteya.net).

[1] This piece was originally published as
Grech Victor. "The Elicitation of Jung's Shadow in Star Trek." NYRSF Review 306: .

Bibliography

"A Fistful of Datas." Dir. Patrick Stewart. *Star Trek: The Next Generation*. Paramount. November 1992.

"Allegiance." Dir. Winrich Kolbe. *Star Trek: The Next Generation*. Paramount. March 1990.

"Amok Time." Dir. Joseph Pevney. *Star Trek: The Original Series*. Paramount. September 1967.

"And the Children Shall Lead." Dir. Marvin Chomsky. *Star Trek: The Original Series*. Paramount. October 1968.

"Battle Lines." Dir. Paul Lynch. *Star Trek: Deep Space Nine*. Paramount. April 1993.

"Brothers." Dir. Rob Bowman. *Star Trek: The Next Generation*. Paramount. October 1990.

"Charlie X." Dir. Lawrence Dobkin. *Star Trek: The Original Series*. Paramount. September 1966.

"Conundrum." Dir. Les Landau. *Star Trek: The Next Generation*. Paramount. February 1992.

"Crossover." Dir. David Livingston. *Star Trek: Deep Space Nine*. Paramount. May 1994.

"Darkling." Dir. Alexander Singer. *Star Trek: Voyager*. Paramount. February 1997.

"Datalore." Dir. Rob Bowman. *Star Trek: The Next Generation*. Paramount. January 1988.

"Day of the Dove." Dir. Marvin Chomsky. *Star Trek: The Original Series*. Paramount. November 1968.

"Deadlock." Dir. David Livingston. *Star Trek: Voyager*. Paramount. March 1996.

"Descent." Dir. Alexander Singer. *Star Trek: The Next Generation*. Paramount. September 1993.

"Distant Voices." Dir. Alexander Singer. *Star Trek: Deep Space Nine*. Paramount. April 1995.

"Emergence." Dir. Cliff Bole. *Star Trek: The Next Generation*. Paramount. May 1994.

"Empok Nor." Dir. Mike Vejar. *Star Trek: Deep Space Nine*. Paramount. May 1997.

"Encounter at Farpoint." Dir. Allen Corey. *Star Trek: The Next Generation*. September 1987.

"Equilibrium." Dir. Cliff Bole. *Star Trek: Deep Space Nine*. Paramount. October 1994.

"Extreme Measures." Dir. Steve Posey. *Star Trek: Deep Space Nine*. Paramount. May 1999.

"Face of the Enemy." Dir. Gabrielle Beaumont. *Star Trek: The Next Generation*. Paramount. February 1993.

"Faces." Dir. Winrich Kolbe. *Star Trek: Voyager*. Paramount. May 1995.

"Facets." Dir. Cliff Bole. *Star Trek: Deep Space Nine*. Paramount. June 1995.

"Frame of Mind." Dir. James L. Conway. *Star Trek: The Next Generation*. Paramount. May 1993.

"Heroes and Demons." Dir. Les Landau. *Star Trek: Voyager*. Paramount. April 1995. "If Wishes Were Horses." Dir. Robert Legato. *Star Trek: Deep Space Nine*. Paramount. May 1993.

"Legacy." Dir. Robert Scheerer. *Star Trek: The Next Generation*. October 1990.

"Man of the People." Dir. Winrich Kolbe. *Star Trek: The Next Generation*. October 1992.

"Mirror, Mirror." Dir. Marc Daniels. *Star Trek: The Original Series*. October 1967.

"Parallels." Dir. Robert Wiemer. *Star Trek: The Next Generation*. November 1993.

"Second Sight." Dir. Alexander Singer. *Star Trek: Deep Space Nine*. Paramount. November 1993.

"Silicon Avatar." Dir. Cliff Bole. *Star Trek: The Next Generation*. Paramount. October 1991.

"Skin of Evil." Dir. Joseph L. Scanlan. *Star Trek: The Next Generation*. April 1988.

"The Alternative Factor." Dir. Gerd Oswald. *Star Trek: The Original Series*. Paramount. March 1967.

"The Chase." Dir. Jonathan Frakes. *Star Trek: The Next Generation*. Paramount. April 1993.

"The Dogs of War." Dir. Avery Brooks. *Star Trek: Deep Space Nine*. Paramount. May 1999.

"The Enemy Within." Dir. Leo Penn. *Star Trek: The Original Series*. Paramount. October 1966.

"The Enterprise Incident." Dir. John Meredyth Lucas. *Star Trek: The Original Series*. Paramount. September 1968.

"The Forge." Dir. Michael Grossman. *Star Trek: Enterprise.* Paramount. November 2004.

"The Return of the Archons." Dir. Joseph Pevney. *Star Trek: The Original Series*. Paramount. February 1967.

"The Siege of AR-558." Dir. Winrich Kolbe. *Star Trek: Deep Space Nine*. Paramount. November 1998.

"The Squire of Gothos." Dir. Don McDougall. *Star Trek: The Original Series*. Paramount. January 1967.

"The Sword of Kahless." Dir. LeVar Burton. *Star Trek: Deep Space Nine*. Paramount. November 1995.

"Time Squared." Dir. Joseph L. Scanlan. *Star Trek: The Next Generation*. Paramount. April 1989.

"True Q." Dir. Robert Scheerer. *Star Trek: The Next Generation*. Paramount. November 1992.

"Violations." Dir. Robert Wiemer. *Star Trek: The Next Generation*. Paramount. February 1992.

"What Are Little Girls Made Of?" Dir. James Goldstone. *Star Trek: The Original Series*. Paramount. October 1966.

"Whispers." Dir. Les Landau. *Star Trek: Deep Space Nine*. Paramount. February 1994.

"Whom Gods Destroy." Dir. Herb Wallerstein. *Star Trek: The Original Series*. Paramount. January 1969.

"Yesterday's Enterprise." Dir. David Carson. *Star Trek: The Next Generation*. February 1990.

Badmington, Neil. *Alien Chic. Posthumanism and the Other Within*. New York: Routledge, 2004.

Barad, Judith and Ed Robertson. *The Ethics of Star Trek*. New York: HarperCollins, 2000.

Blair Karin. "The Garden in the Machine: The Why of Star Trek." *The Journal of Popular Culture*, 13.2 (1979): 310–320.

Blish, James. *Spock Must Die!* New York: Bantam, 1970.

Eberl, Jason T. and Kevin S. Decker. *Star Trek and Philosophy: The Wrath of Kant*. Chicago: Open Court, 2008.

Everett, Hugh III. "The Theory of the Universal Wavefunction" Diss. Princeton, 1957

Grech Victor. "The Irruption of Vulcan Pon Farr as Unleashment of Jung's Shadow." *New York Review of Science Fiction*, 25 (2012): 13-15.

Grech, Victor. "The Pinocchio Syndrome and the Prosthetic Impulse in Science Fiction." *New York Review of Science Fiction*, 24.284 (2012): 11-15.

Grech, Victor. "The Trick of Hard SF is to Minimize Cheating Not Just Disguise It With Fancy Footwork: The Transporter in Star Trek: Can It Work?" *Foundation*, 111 (2011): 52-67.

Hart, David L. "The Classical Jungian School." *The Cambridge Companion to Jung*. Eds. Polly Young-Eisendrath and Terence Dawson. Cambridge: Cambridge UP, 1997.

Jung, Carl Gustav. "A Psychological Approach to the Dogma of the Trinity." *Psychology and Religion: West and East. Collected Works.* Vol. 11. 1938. New York: Pantheon, 1957.

Jung, Carl Gustav. "Analytical Psychology and Weltanschauung." *The Structure and Dynamics of the Psyche. Collected Works.* Vol. 8. 1928. New York: Pantheon, 1957.

Jung, Carl Gustav. "Answer to Job." *Psychology and Religion: West and East. Collected Works.* Vol. 11. 1938. New York: Pantheon, 1957.

Jung, Carl Gustav. "New Paths in Psychology." *Two Essays on Analytical Psychology. Collected Works.* Vol. 7. 1912. New York: Pantheon, 1957.

Jung, Carl Gustav. "On the Psychology of the Unconscious." *Two Essays on Analytical Psychology. Collected Works.* Vol. 7. 1912. New York: Pantheon, 1957.

Jung, Carl Gustav. "Psychology and Religion." *Psychology and Religion: West and East. Collected Works.* Vol. 11. 1938. New York: Pantheon, 1957.

Jung, Carl Gustav. "Psychotherapist or the Clergy." *Psychology and Religion: West and East. Collected Works.* Vol. 11. 1938. New York: Pantheon, 1957.

Jung, Carl Gustav. "The Psychology of the Child Archetype." *The Archetypes and the Collective Unconscious. Collected Works.* Vol. 9. 1940. New York: Pantheon, 1957.

Jung, Carl Gustav. *Aion. Collected Works.* Vol. 9. New York: Pantheon, 1951.

Jung, Carl Gustav. *Memories, Dreams, Reflections.* 1963. London: Flamingo, 1983.

Jung, Carl Gustav. *MysteriumConiunctionis: Inquiry into the Separation and Synthesis of Psychic Opposites in*

Alchemy. Collected Works. Vol. 14. 1963. London: Routledge, 1963.

Kapell, Matthew Wilhelm. *Star Trek as Myth. Essays on Symbol and Archetype at the Final Frontier.* London: McFarland & Company, Inc., 2010.

Lundeen, Jan and Jon Wagner. *Deep Space and Sacred Time: Star Trek in the American Mythos.* Westport: Praegher, 1998.

Nietzsche, Friedrich Wilhelm. *Beyond Good and Evil.* Trans. Helen Zimmern. 1886. Rockville: Serenity, 2012.

Rogers, Ben, Robert Wagner, Qichang Su and Rainer Grobe. "Reconstruction of Objects in Random edia Based on Their Shadow Patterns." *Bulletin of the American Physical Society* 56.13 (2011):1

Rogers, Robert. *A Psychoanalytic Study of the Double in Literature.* Detroit: Wayne State University Press, 1970.

Schelde, Per. *Androids, Humanoids, and Other Science Fiction Monsters : Science and Soul in Science Fiction Films.* New York: New York UP, 1993.

Star Trek IV: The Voyage Home. Dir. Leonard Nimoy. Paramount. 1986.

Star Trek Nemesis. Dir. Baird Stuart. Paramount Pictures, 2002.

Star Trek V: The Final Frontier. Dir. William Shatner. Paramount. 1989.

Stevenson Robert Louis. *The Strange Case of Dr. Jekyll and Mr. Hyde.* London: Longmans, Green & Co., 1886.

Chapter 4.   The Roles of Women in Murray Leinster's *Med Ship* Stories[2] - Mariella Scerri and Victor Grech

Introduction

The role of women in literature has been a source of constant debate since the early twentieth century. As far back as the Victorian era, female authors strove to establish their identity, both in literature and in life. According to the standard narrative of feminist intellectual history, modern feminism in the English-speaking world begins with Mary Wollstonecraft's bold appeals for women's inclusion in a public life overwhelmingly   dominated   by   men (Wollstonecraft). So much so that in a literary world dominated by men, she sought to retain her maiden surname in her oeuvre, marking her as a fierce and independent woman. Specific attention is drawn to her theories of character formation and the importance of public education for women in order to foster their cognitive abilities.

Wollstonecraft's *A Vindication of the Rights of Woman* is a celebration of the rationality of women (Wollstonecraft). It constitutes an attack on the view of female education put forward by Rousseau and

---

[2] This piece was originally published in the NYRSF Review as Scerri, Mariella and Grech, Victor. "The Roles of Women in Murray Leinster's *Med Ship* Stories." *The New York Review of Science Fiction.* 317 (2015): 28-30.

countless others who regarded women as weak and not capable of effective reasoning. The state of degradation was considered as the feminine norm as accepted by their male counterparts during Wollstonecraft's era, the position to which women are consigned through the designations of the patriarchal male.

The idea of woman as alien in a patriarchal culture was depicted and accepted not only from men and also from women. Virginia Woolf, a major twentieth century novelist and one of the foremost modernists and pioneer of feminist criticism famously stated "A woman must have money and a room of her own if she is to write fiction" (Woolf). Woolf's best known nonfiction works, *A Room of One's Own* (1929) and *Three Guineas* (1938), examined the difficulties that female writers and intellectuals faced because men held disproportionate legal and economic power. Woolf's *Three Guineas* is exceedingly complex and is exasperatingly persuasive in bringing into focus the marginal experience of women in a patriarchal society. Her entire purpose in *Three Guineas* was not to create something out of nothing, or spontaneously generate facts, but to bring what already is seen but marginal to the foreground to show the inherent connections that link the lesser concern to the greater situation. Furthermore, her information was entirely based upon press clippings and secondary sources such as biographies, presenting an accurate assessment and factual information of a situation connected to a serious problem. If anything she attempted to fight against the propaganda of patriarchy in order to reveal the "false truth" (Park).

Feminist Literary Criticism

Woolf's reputation declined sharply after World War II, but her importance was re-established with the growth of feminist criticism in the 1970's. Feminist critics have made a massive contribution to challenging the notion of a received literary canon inscribed by male authors. Up to the first half of the twentieth century, readers followed the assumptions dictated by the male patriarchy and rarely challenged the status quo. Simone De Beauvoir, whose monumental *The Second Sex* (1949), was published twenty years after Woolf's *A Room of One's Own*, proposed a radical approach. Accompanying her exploration of the role of women in literature by scrutinizing their place in biology, anthropology, religion and philosophy, her Introduction brings out her central argument: that women's role has been socially constructed defined in relation to men- who are seen as the "Absolute", while women are "the Other" (De Beauvoir).

French critic Helene Cixous, proclaims the possibility of a feminine writing – *ecriture feminine* – which would break down the barriers excluding women from public speech. Indeed, Cixous asserted: *"We the precocious, we the repressed of culture, our lovely mouths gagged with pollen, our wind knocked out of us, we the labyrinths, the ladders, the trampled spaces, the bevies – we are black and we are beautiful"* (247).

Cixous's argument converges with Jean-Paul Sartre's point of departure when he claims that "the committed writer knows that words are action." From this point of view we may conclude that the writer has chosen to reveal the world and particularly to reveal man to other men, so that the latter "may assume full responsibility before the object which has been laid bare". Thus to Sartre, "people of the same period and community" of the author "have the same complicity" (Sartre 12-15).

Gender in speculative fiction

This leads to the question: What is the role of women in an avant-garde genre such as science fiction. Gender has been an important theme that has been widely explored in speculative fiction. The genres that make up speculative fiction, science fiction (SF), fantasy, supernatural horror and related genres have always offered the opportunity for writers to explore social conventions, including gender, gender roles and beliefs about gender. Many writers have chosen to write with little or no questioning of gender roles instead effectively reflecting their own cultural gender roles onto their fictional world. Like all literary forms, the science fiction genre reflects the contemporary ideas and conventions which individual authors were writing, thus grounding those authors' responses to gender stereotypes and gender roles.

Science fiction in particular has traditionally been a "puritanical genre orientated toward a male readership" (Clute and Nicholls 1343). Most of the "stereotypical tropes of science fiction, such as aliens, robots or superpowers can be employed in such a way as to be metaphors for gender" (Attebery 1).

Many male protagonists of science fiction are reflections of a single heroic archetype, often having scientific vocations or interests and being "cool, rational, competent", remarkably sexless, "interchangeable and bland" (Kuhn) as long as he serves the role of the Campbellian hero (Campbell). On the other hand, the common perception of the role of women in SF works has long been dominated by one of two stereotypes: a woman who is evil (villainess) or one who is helpless (damsel in distress).

The minimal role of women in Med Ship

These above arguments provide an analytic framework for the discussion of the microscopic role of women in Leinster's *Med Ship*. Murray Leinster (June 6, 1890 – June 8, 1975) was a *nom de plume* of William Fitzgerald Jenkins, an award winning American writer of science fiction and alternative history. *Med Ship* is a collection of stories written by Murray Leinster between 1957 and 1966. They are the continuing adventures of Calhoun, a physician with the "Interstellar Medical Service." He travels the galaxy in his own small spaceship with his pet and companion Murgatroyd, visiting planets to try to solve public health issues. Each story is self contained, and tells Calhoun's adventures on a new planet that has a health crisis. Similar to *Medicins Sans Frontieres* International, *Med Ship* volunteers doctors who travel from world to world with no actual enforcement powers but who are so respected that their advice is never questioned (Grech 6).

All eight stories involve male dominated worlds and focus exclusively on men. Written in the mid 20th century, before the rise of the feminist movement, it can be assumed, that *Med Ship*'s target audience is predominantly male. Science fiction has traditionally been viewed as a male-oriented genre and originally had a reputation of being created by men for other men (Tuttle). All the stories involved in *Med Ship* rely on a substantial amount of descriptive technical information, which might draw away the female reader without a scientific background. For example, Calhoun's natural choice for help to assist him in disabling a ground-induction field in "Med Ship Man" (1963) was a troop of men. They are described as "able-bodies and grim-faced men. Two were electronic engineers, as he'd specified. One was a policeman. There were two mechanics and a doctor..." (Leinster 41). On "Tallien III" (1963) the reader encounters a mass population of men who are either normal or "paras." While it gives the impression of a heavily populated city, with a structural framework of a civilized society, there is no mention of either women or children.

The first female is introduced half way through the third story "The Mutant Weapon" (1959) and is presented to us as an emaciated girl. "A girl emerged from the thicket. She was gaunt and thin, yet her garments had once been of admirable quality" (Leinster 147). The hunger-stricken girl immediately brings to mind the Victorian image of the sickly girl. Gilbert and Gubar claim that diseases of maladjustment to the physical and social environment as anorexia and agoraphobia do strike a disproportionate number of women. Such diseases are caused by patriarchal socialization in several ways. In the nineteenth century, the desire to be beautiful and frail was predominant (Gilbert and Gubar). The portrayal of the girl in "Mutant

Weapon," (1959) thus parallels other Victorian women and is also enforced in the science fiction genre. Further, her first reaction to "murder him from ambush" is the kind of resistance associated with the fight against submissiveness to male dominance and patriarchy. The notion of the beautiful image of young women is perpetuated further on in same story. The omnisicient narrator claims, "There are people who, because they are physically unattractive, become personalities. All too many girls – and men, too – do not bother to become anything but good to look at" (Leinster 204). It is worth noting here the hesitant claim "– and men, too" which provides a weak and uncertain affirmation of the inclusion of men to be held as objects of desire by the society at large.

In addition, the minor role given to this girl is that of an assistant to her medical boyfriend, which such position was later relegated to help Calhoun. Regardless of their intelligence or capabilities, women in *Med Ship* are given inferior positions and scorned for their weaknesses by explicit remarks, implicit subtle innuendos or worse still by complete omission.

Where women are mentioned in *Med Ship*'s short stories, they are always implicated with a lover and an intriguing love story. They never stand alone. In "Ribbon in the Sky" (1957), Leinster portrays three belligerent cities which constantly battle each other. They forbid association from one city to another due to an irrational fear of contagion of a plague. The story revolves around a young girl and a young man terribly in love, in the trope of Romeo and Juliet. "Some young girl must have loved terribly, and some young man been no less impassioned to accept expulsion from society on a world where there was no food except in hydroponic gardens and artificially warmed pastures. It was no less suicide for those who loved" (Leinster 260). The breach of law of these young adults condemned them to abandonment by their families and death in the "hotlands". Only after Hunt (the girl's father) accepted Calhoun's rational explanation with regard to the absence of any plague, did he accept the marriage proposal and the ensuing intermingling of the two cities. The resolution of the situation in marriage echoes the denouement in fairytales and would have been unacceptable in realist novels. But the happy ending of most science fiction books and films, makes the ending of "Ribbon in the Sky"(1957) more welcome and acceptable. Infact, Grech et al claim there is an almost universal outcome in the science fiction genre, in that a solution is always found, one that restores situation to normality, thus "underscoring the genre's penchant for happy endings" (Grech et al 27).

The nurturing role of women predominates in "Grandfather's Wars" (1957), where girls as a collective group are seen tending for younger children on Canis III, sent from their planet of origin Phaedra because of the danger resulting from the instability of the planet's sun. The girls' lack of know-how in raising their younger generation is apparent. But what is even more obvious is the naturalisation of gender roles assumed in young adults. For the young men, "the instinct of their age group directed them as specifically as successive generations of social insects are directed. They moved about in gangs.... the warrior age group would be capable of immense and admirable skill in handling anything which interested them..." For the young women, "Deep-rooted instincts still worked. Women – young women – and girls appeared still to feel concern for young children which were not even their own" (Leinster 464). The stereotypic primordial roles of both the girls and the young men occur again toward the end of the story, where the young men "...have taken to the woods. They swear they "never give in!"", while the girls are "... fluttering about and beginning to talk about clothes. When older women arrive "there'll be dress making" (Leinster 500).

"Quarantine World" (1966) and "Pariah Planet" (1961) are the only two stories in *Med Ship* which allow women any sort of significant role. The girl in "Quarantine World" (1966) is brought by the narrator into action in a nurturing role. "He saw, with clearing eyes that a figure bent over him. It was a girl with dark brown eyes. She lifted his head and gave him a drink from a cup" (Leinster 387). But she soon turns into a good source of information for Calhoun, describing the situation on her planet in great detail and answering all of his queries. It is the first female role given credence and status thus far in the book. The girl Elna provides information to Calhoun regarding the fraught situation between her planet Delhi and Lanke and the predominant and irrational fear of contagion of a plague which exists solely on Delhi, of which the natives seem to be immune. The fact that the girl is even given a name in this story continues to show her importance.

The autonomy of the girl Elna is however sabotaged by Rob, who though physical aggressiveness demonstrates patriarchal domination. "Rob said in icy fury, "You're a woman and I'd have had to hurt you to keep you from interfering. It's because you've been listening to him! ..." (Leinster 404). The reader thus witnesses the two competing men, rivals to each other in the bigger picture – the situation on Delhi and Calhoun's determination to resolve it. Calhoun here mirrors the single heroic archetype – cool, rational, competent and victorious. "He drew the pocket-blaster from under his robe" (Leinster 405).

The last story "Pariah Planet" (1961) is the second and last story which gives women recognition, and provides a welcome exception from the stereotypic role usually associated with the book's previous women of nurturing and love. Maril is intelligent, quick to learn and a good assistant to the Medical Service man Calhoun. But Maril is still regarded as man's inferior by Calhoun himself. Being in control in each and every situation, it can hardly be assumed that Calhoun will relinquish his control. His patronizing streak may even go unnoticed by his reader audience. "Good girl! He said approvingly. "I'll give this back to you when we land" (Leinster 536), he claimed when she tried to steal his blaster from him.

Maril's resilience and refusal to talk makes her an antithetical definition to the widely claimed chattering of women. Calhoun has to arrive to his own conclusions regarding the girl's past. Her resistance and continuing fight against Calhoun's subordination leads her to escape once they land on Dara – an idiotic move which almost leads to her death. Once again, the reader recognizes the archetypal hero in Calhoun, who saves his conquest from an animal stampede out of sheer luck. Moments later he rescued her once again, when three Dara men opened fire on them. "He jerked the girl Maril to her feet and rushed her toward the Med Ship. Smoke from the flung bomb upwind barely swirled around him and missed Maril altogether" (Leinster 553). The subtle innuendos about the inferiority of women reach their climax, with a derogatory comment from the narrator. "He was a professional man. In his profession he was not incompetent. But there is no profession in which a really competent man tries to understand woman" (Leinster 564).

The inferiority with which the book treats women is further pronounced when Calhoun decides to train two groups of pilots to steal spaceships filled with provisions floating in orbit and navigate them back to their home planet. Maril hovers in servitude on Med Ship, a menial role relegated to the inferior gender – the "Other." The heroic role is appropriated by the "Absolute" males (De Beauvoir). Upon expressing her wish of piloting a ship, Calhoun answered, "You wouldn't want to be a heroine. No normal girl does." Calhoun goes further in saying, "[Korvan] wouldn't feel comfortable with a girl who'd helped make starving unnecessary. He'd admire you politely, but he'd never marry you. And you know it" (Leinster 594).

Discussion

The study of women within science fiction in the last decades of the 20th century has been driven in part by the feminist and gay liberation movements, and has included strands of the variously related and spin off movements, such as gender studies and queer theory. The portrayal of women, or more broadly, the portrayal of gender in science fiction, has fluctuated throughout the genre's history. Some writers and artists have challenged their society's gender norms in producing their work. Others have not. Written before the feminist movement, *Med Ship* does not challenge role expectations. It affirms Garber and Paleo's statement that "female characters were only occasionally included in science fiction pulp stories" (Garber and Paleo). Their inclusion only reinforces their status quo of submissiveness, deemed as the "Other" by the males "Absolute" who subscribe to it, in this case the omniscient narrator and the hero Calhoun.

Enjoying the accolade as the "dean of science fiction," Murray Leinster was a pioneer in this genre and successfully established the sub genre of the "science fiction doctor story" (Flint and Gordon). However, through his treatment of women roles, he followed contemporary, conventional ideas. The common tropes of love and nurture are predominant in the few instances where women are portrayed. There is an implicit consensus throughout *Med Ship's* narrative of a male supremacy across various planets of the universe, while women are repetitively considered as weaklings to be ignored. This narrative thus parallels other contemporary fiction books written by male authors. The patriarchal stronghold was still unshakeable until late 20th century, and men both in literature and in society were still enjoying the patriarchal domination and the inherent advantages which such rule brings along. The Victorian image of the female role as wife, mother and housekeeper, although challenged, was perpetuated into modernism and other related genres including science fiction.

After the 1970s, science fiction saw a dramatic change. Several events began to focus on women in fandom, professional science fiction and as characters. In 1974, Pamela Sargent published an influential anthology, *Women of* Science *Wonder: Fiction Stories by Women, about Women*, the first of many anthologies to come that focused on women or gender rules. Additionally, movement among writers concerned with feminism and gender roles sprang up, leading to a genre of feminist science fiction including Joanna Russ's, *The Female Man* (1978), Samuel R. Delany's 1976, *Trouble on Triton: An Ambiguous Heterotopia*, and Marge Piercy's 1976, *Woman on the Edge of Time*. These authors began to explore science fiction as an alternative to the realism and realist conventions espoused by many feminists. Since the 1970s feminist writers and critics like Ursula K. Le Guin and Joanna Russ have commented on Western patriarchal capitalism through science fiction. At the same time feminist critics like Donna Haraway hailed feminist science fiction as one of the most productive sites for imagining better "social arrangements and theorizing our way out of a constructing humanism" (Jessert). Russ cautiously embraces science and technology as a place for feminist interventions in patriarchal capitalism.

Reading through this collection it becomes clear that feminist studies of science fiction were shaped on the margins of patriarchal rule to gain entrance and legitimacy. Such entry from a marginalized position parallels other attempts by women writers into male dominated genres and has come full circle, when it comes to women's writing regardless of the era in which it was written or what type of genre.

References

Attebery Brian. *Decoding Gender in Science Fiction*. New York: Routledge, 2002.

Cixous Helene. *The Laugh of the Medusa*. Brighton: Harvester Press, 1981.

Campbell Joseph. *The Hero with a Thousand Faces*. USA: Pantheon Books, 1949.

Clute John and Nicholls Peter. *The Encyclopedia of Science Fiction*. Great Britain: Orbit, 1999.

De Beauvoir Simone. Introduction, *The Second Sex*. Great Britain: Pan Books, 1988.

Delaney Samuel R. *Trouble on Triton*. United States: Bantam Books, 1976.

Flint Eric and Gordon Guy (eds). *Med Ship*. USA: Baen Publishing Enterprises, 2002.

Gilbert Sandra M. and Gubar Susan. *The Madwoman in the Attic*. USA: Library of Congress Cataloging in Publication Data, 1979.

Grech Victor. "Doctor by Doctor: Dr. Philip Boyce and Dr. Mark Piper in Star Trek..." *Vector* Winter 274 (2103/14): 6-9.

Grech Victor, Vassallo Clare and Callus Ivan. "The Last (fertile) Man on Earth: Comedy or Fantasy? *World Future Review* (2013) 5(24): 24-28

Jessert Nancy. *Dreams worth watching? Science fiction and the futures of feminism*. USA: MPublishing, University of Michigan Library, 1997.

Kuhn Annette. Alien Zone: *Cultural Theory and Contemporary Science Fiction Cinema*. London: Verso, 1990.

Leinster Murray. (eds Flint Eric and Gordon Guy). *Med Ship*. USA: Baen Publishing Enterprises, 2002.

Park Sowon S. "Suffrage and Virginia Woolf: The Mass Behind the Single Voice." *The Review of English Studies, New Series* February (2005) 56 (23): 119-134.

Piercy Marge. *Woman on the Edge of Time*. USA: Alfred A. Knopf, 1976.

Russ Joanna. *The Female Man*. USA: Bantam Books, 1975.

Sargent Pamela. *Women of Science Wonder*. USA: Vintage Books, 1996.

Sartre Jean Paul. *What is literature*. United Kingdom: Metheun, 1967.

Tuttle Lisa. "Women as portrayed in Science Fiction." *The Encyclopedia of Science Fiction*. United Kingdom: Granada, 1979.

Wollstonecraft Mary. *A Vindication of the Rights of Woman*. Boston: Peter Edes, 1792.

Woolf Virginia. *A Room of One's Own*. United Kingdom: Hogarth Press, 1929.

Woolf Virginia. *Three Guineas*. United Kingdom : Hogarth Press, 1938.

Chapter 5. Infertility in Science Fiction as a Feminist Issue – Clare Vassallo and Victor Grech

Introduction: Feminism and Fertility

Although mythological figures such as the Amazonian female warrior might encourage us to think that 'feminism is as old as mythology' (Hard 2004 263), feminism as a political stance through which the personal came to be perceived as political and which highlighted patriarchal structures and ideology as systematically making the female of the human species seem inferior to the male, is a relatively recent phenomenon. Despite some important 19th century works, we can claim that feminism as a political and critical movement came into its own after WWII.

Literary feminism brings together a range of approaches to textual analysis. These include the critique of patriarchal language and tropes, addressing the historical disappearance of women writers, the authorial voice of women, and the increasing presence of women in the canon and in genres formally dominated by male writers. Some of these approaches, such as Marxist Feminism, have privileged the realist novel as a locus of analysis and have delved into the relationship between literature and life, perceiving representation of women, attitudes towards women, and the language of the text as reflective of socially constructed gender-biased attitudes between the sexes at various historical moments (Morris 1993; Eagleton 1996; Lefanu 1995; Russ 1995).

The sub-genre of feminist science fiction with an emphasis on fertility explores the roles of women and men by examining social constructions and the enforcement of gender roles with particular reference to the inequalities of personal and political power that are dictated by one's gender. Feminist SF often delves into these themes by contrasting two opposed approaches: utopias and dystopias. The former tends towards narrative worlds in which gender differences are non-existent, as in single-sex worlds, whereas dystopias tend towards the extrapolation of patriarchal structures taken to extremes. Since SF is genre closely connected to realist social concerns, its critical and innovative centre is more likely to be found in its themes and content rather than in its formal structures, which tend to follow fairly conventional narratological patterns.

Feminism and fertility are inextricably intertwined. Varying viewpoints on the relationship between the political and the physical can be plotted in a typology of different feminisms with an eye to fertility and related issues. Broad feminist categories can be outlined along the following axis:

Liberal feminism identifies its roots with Mary Wollstonecraft and John Stuart Mill, seeking equality of the two sexes due to the fear that the different treatment of women that may lead to the stereotyping of roles and to marginalization, typically through marriage. Pregnancy is viewed as a disability, while new reproductive technologies are welcomed as augmenting female choice. Men are therefore involved as in statements such as 'we are pregnant' but this, on the other hand, implies male appropriation of the female body.

Matriarchal/matrifocal feminism revels in the ways in which women differ from men and therefore celebrates pregnancy as an identity-conferring condition. Infertility treatments are viewed with suspicion as they are seen as patriarchal techniques that medicalize and dominate the body (Corea 1985). Interestingly, Peggy Robin points out that 85% of women seeking infertility treatment were attended by male physicians (Robin 1993).

Postmodern feminists are arguably the least concerned with these issues by regarding 'femaleness' as a socially constructed category, they study the ways in which class, race, and other factors, such as gender, construct the female 'body'. *French Feminism* is also less concerned with the body and more focussed on issues of style and voice in writing, indicating that a feminine style does not necessarily indicate a female writer.

Foucauldian feminists view the most dangerous forms of control as those that are ubiquitous and so pervasive that they are assumed and deeply internalized such that these strictures appear disconnected from any form of overt displays of power, and indeed, may even disguise themselves as forms of liberation and choice. This may include relatively innocuous and ostensibly helpful state surrogates, such as home health visitors after delivery, as the state has a stake in the health of both mother and child, with confinement here having a double meaning.

Feminism and SF

The 1970s was the decade in which feminism in science fiction flourished. This is the genre that, arguably, produced some of the most interesting examples of feminist fiction. Carl Freedman's review of Marleen S. Barr's *Future Females: The Next Generation* (2000) brings into steady focus the fact that the intertwining of feminism and science fiction was flourishing long before the literary institutions and feminist critics paid it any attention. The popular, rather than academic, status long attributed to SF meant that works which today are considered central to the symbiosis of feminism and

SF were ignored for at least a decade after their publication.

In an ironic development, it was the women writers of SF who were instrumental in bringing SF to academic critical recognition. As these new writers of SF brought mainstream themes of race, gender and class, elaborated through the canonical fictional devices of utopias and dystopias through a feminist point of view to the genre their works began to impinge on critical notice. The genre, which was previously associated mainly with interplanetary warfare, medieval looking worlds with anachronistic weapons set in future time/space scenarios and which seemed to have little relevance to the present conditions of humanity and of writing, was suddenly involved with the same themes that more 'literary' texts were developing.

In addition, as writers who had already received critical attention through non-SF novels, such as Ursula K. Le Guin, known for her work in fantasy, and Margaret Atwood, well-known for award winning novels which were fully based in the 'real' world, began to write SF the attention of the academy was engaged. These 'cross-over' writers were instrumental in drawing the critical gaze to the genre of science fiction.

Utopian and Dystopian Narratives

Two of the most notable novels detailing feminist utopias are Le Guin's *Left Hand of Darkness* (1969) 'which, through a radical imagining of human life without gender, explores gender as a cultural construction that is at once coercive and contingent' (Pearson et al. 2010 5), and Russ's *The Female Man* (1975) 'which focuses on the struggle to establish lesbian and feminist identities and sexualities within the constraints of a culture of compulsory heterosexuality' (Pearson et al. 2010 5). Both are novels of the golden age of 70s feminist SF. In *The Female Man*, 'characters refuse the reader's search for innocent wholeness while granting the wish for heroic quests, exuberant eroticism, and serious politics'. The book deals with four women who hail from different worlds: Jeannine whose world revolves around marriage, Joanna who is experiencing a feminist revolution but is still expected to orient herself around men, Janet who lives in a women-only world as men have been killed off centuries before by a plague, and Jael, an assassin who lives in a world where the two sexes wage a cold war. These individuals are 'four versions of one genotype, all of whom meet, but even taken together do not make a whole, resolve the dilemmas of violent moral action, or remove the growing scandal of gender'(Haraway 1991 178).

Feminist dystopias create societies wherein gender inequities are actually exaggerated and intensified, and perhaps Atwood's *The Handmaid's Tale* (1986) quintessentially embodies the ultimate of such possible dystopias. Inevitably, feminist dystopian fiction has also described a turning of the tables as in the 1905 short story, *The Sultana's Dream,* by Rokeya Sakhawat Hussain which portrays an alternate, crime-free world where men exist in a state of gender-reversed purdah.

Vonda McIntyre visualises fertility as a feminist contraceptive issue, and in *Dreamsnake* (1978) and *Superluminal* (1984), she visualises both male and female fertility as a voluntary individual decision through a form of auto-control. The latter helps 'redefine the pleasures and politics of embodiment and feminist writing. In a fiction where no character is 'simply' human, human status is highly problematic', as humanity is transformed by 'bionic implants, […] virus vectors carrying a new developmental code, by transplant surgery, by implants of microelectronic devices, by analogue doubles, and other means' (Haraway 1991 180).

In Frank Herbert's magnum opus *Dune* (1965), female fertility control is taken even further through the semi-religious, hereditary, female-only group known as the 'Bene Gesserit' who can also determine the gender of their offspring. They also play a byzantine game of politics through a breeding program spanning a millennium in order to produce a hyper-evolved but male mental adept. *Dune* also portrays a member of the nobility who is an offshoot of this breeding program and who is described as being a genetic eunuch, presumably implying a male who is born sterile. It is worth mentioning at this point that the long-term breeding of various species, both humans on Earth and aliens on other planets, is also the main theme behind what is probably the most famous space opera of all time, Smith's *Lensman* series. In true Stapledonian fashion, the story begins two thousand million years prior to contemporary events, when a benevolent alien race commences the breeding of several intelligent species in order to hand over the guardianship of the very universe in the face of an implacable and evil alien invasion of the universe (Smith 1948). An unresolved plot element in the last book of the series concerns the ultimate development of this breeding program, five children, four of whom are women, who cannot possibly find anyone interesting enough to mate with, potentially resulting in their infertility (Smith 1948).

Yet another aspect of extreme physiological control of pregnancy is depicted in Iain M. Bank's *Excession* (1996), set in the *Culture* universe, where a human female carries a deliberately arrested gestation for years. Interestingly, female sexuality has also been utilised as a form of projectile weapon in Barker's

SF-fantasy pastiche *Weaveworld* (1987). Barker describes an ancient humanity that could access magic, but as science asserted its dominance, these individuals retreated to secret hideaways. A female renegade uses her magical powers, including the 'menstruum', to attempt to rule the seerkind or destroy them. 'Menstruum' was actually a Latin word for a solvent, which was specifically used in extracting compounds from plant and animal tissues for the preparation of drugs. In a more practical vein, in Bujold's *Vorkosigan* universe, women who have had a contraceptive implant wear a distinguishing earring to state that they are consenting and contraceptive-protected adults.

Totalitarian Ideology and Fertility
The political vision of dystopian literature is often linked to totalitarian regimes in which the individual becomes a pawn of the state or of a ruling elite, and demographics an issue to be decided and determined by the state rather than the individual. Margaret Atwood highlights the politics of female fertility in the fictional world of *The Handmaid's Tale* by typically linking real world events to the fictionalized future scenario. In her 2001 publication *In Other Worlds: SF and the Human Imagination*, she explains:

> My rules for *The Handmaid's Tale* were simple: I would not put into this book

anything that humankind had not already done, somewhere, sometime, or for which it did not already have the tools (Atwood 2001: 88).

One of the reasons that feminism has flourished in SF, as previously mentioned, is the possibility of imaging worlds in which gender constructed differences cease to exist, sometimes, as in the case of single gendered imaginary worlds, because there is only one type of human. Fictional utopias provide the freedom within which to construct, to use Le Guin's term, 'thought experiments' that can play with combinations or absences that would blend or remove all gender imbalance and unfairness as perceived by contemporary women and men readers. Dystopias, on the other hand, provide the impetus to project possible outcomes that can be more harmful and more restrictive for women in different scenarios. Politically, these scenarios tend to be totalitarian regimes of different kinds, including the conservative Puritan theocracy imagined as the future for the US as conceived by Atwood. Such a society might revoke the liberation from gender expectations which contemporary women have begun to take for granted. Atwood asks:

How thin is the ice on which supposedly "liberated" modern Western women stand?

How far can they go? How much trouble are they in? What's down there is they fall? And further: If you were attempting a totalitarian takeover of the United States, how would you do it? What form would such a government assume, and what flag would it fly? How much social instability would it take before people would renounce their hard-won civil liberties in a tradeoff for "safety"? And, since most totalitarianisms we know have attempted to control reproduction one way or another – limiting births, demanding births, specifying who can marry who and who owns the kids – how would that motif play off for women? (Atwood 2001: 87).

Both utopian and dystopian fiction, particularly in the case of SF, provide the author with a construct through which to critique the present, while proposing a possible prediction of a future. Just as one of the essential generic features of 'hard' SF is that the science in the worlds is 'real' science, known in the present and extrapolated into the future as in the case of teleporting humans, portals, space travel, and so on, then one of the generic features of utopian/dystopian fiction is the similar device of taking a trend in the present and 'arriving via logic at a prophetic truth' (Atwood 2001 122). The truth arrived at in a manner typical of the novelist, whose 'business,' Le Guin reminds us, 'is lying.' Precisely the definition of the poetic that Aristotle provided in the earliest discussion of genre, the *Poetics*, in which he separated the writing of history from the writing of poetry precisely on the point that poetry can provide us with the greater truth of the possible ways things can happen, whereas history only speaks of the particulars of what has been. 'Poetry,' he says, 'is a more philosophical and serious business than history; for poetry speaks of universals, history of particulars' (Else 1994 133).

Atwood was called to defend her tale of forced fertility based on biblical ritual as a school text book. She wrote a letter to schools in which she explained,

> The sexual point in my book would seem to be that all totalitarianisms try to control sex and reproduction one way or another. Many have forbidden inter-racial and inter-class unions. Some have tried to limit childbirth, other have tried to enforce it. It was a common practice for slave owners to rape their slaves, for the simple purpose of making more slaves. And so on. (Atwood 2001 244).

In Atwood we seem to have a combination of history and fiction, which gives rise to a chillingly possible future for the West, and in some other parts of the world a reality in the present, in which women's control over their bodies and their fertility is denied them in the name of patriarchal structures and religious norms.

Cyborgs and Feminist SF

Female sexual fulfilment is also explored in feminist narratives such as Piercy's *He, She, and It* (1991), where an android, a re-creation of the equivalent of a Golem by two Jewish scientists, becomes a being who 'transgresses not only the conventional boundary between human and machine, but between male and female as well.' His programming is such that he 'derives his pleasure primarily from pleasing his partner,' a being whose 'marvellous organ is scrupulously clean.' His 'entire body is free of the kind of physical imperfections that characterize human men.' However, this android 'differs substantially from Haraway's notion that the problematic gender of the cyborg is considerably more "dangerous" than that of the sensitive male, whose very androgyny may in fact involve an attempt subtly to appropriate power,' and also imbricates the trope of the sanitisation of sex, a common element in cyberpunk with its technological appropriation and misappropriation, 'a phenomenon embodied, for example, in the distaste for 'meat things' shown by many of Gibson's male characters (Booker 1994).

The converse is Asimov's *Satisfaction Guaranteed* (1951), a short robot story which depicts an experimental humaniform household robot that is fashioned in a very handsome male form. The robot is placed with a woman whose husband works for the robot's manufacturing company. The robot comes to the realisation that she has low self-esteem and attempts to redress this by redecorating his mistress' house and by giving her a make-over through the use of cosmetics and other artifices. At the end of the story, he deliberately allows her neighbours to see him, a strange and handsome male, him kiss her, thereby elevating her social status.

In both of the above stories, sex of any kind with any sort of robot, will naturally not result in pregnancy.

Women-Only Worlds
Women-only worlds abound, and may be viewed as extreme feminist utopias. Only three famous narratives will be highlighted as examples of this subtrope.

Tiptree's most famous award winning and reprinted story is *Houston, Houston, Do You Read?* (1976), and this depicts a plague that wipes out most human life, with only 11,000 people survivors, all female, who continued the species by repeatedly cloning these original 11,000 genotypes. A group of males who return to Earth from space are killed so as to avoid disturbing the harmonious paradise that Earth has become in the absence of the male of the species. Similarly, Russ's *When It Changed* (1972) depicts the return of males to the women-only world of *Whileaway*. This society is stable and peaceful and women see the return of the men as a return to tyranny and oppression of the past, and yet, men assume that they will be eventually made welcome, even if their return is forcefully imposed.

One of the more recent, women-only worlds has been described by Doris Lessing, and in *The Cleft* (2007), an ancient community of women have no knowledge of men, and childbirth is regulated by the cycles of the moon. This feminist utopia is disrupted by the birth of boys.

Discussion
As we have shown above, the single-gendered trope is often used to explore utopian (usually feminist) scenarios or dystopias. Power is enmeshed in all of these discourses, whether feminist or otherwise, as argued by Foucault: '(i) that power is co-extensive with the social body; there are no spaces of primal liberty between the meshes of its network; (ii) that relations of power are interwoven with other kinds of relations (production, kinship, family, sexuality)' (Foucault 1980 142). This is particularly so in sexual relations wherein interpersonal relationships achieve

greatest intricacy and intensity, and are hence particularly susceptible to the mechanisms of power. Fertility and reproduction play key roles in defining gender, and the control of one's fertility is a central theme in feminist manifestos, as pregnancy and childrearing are often used to subordinate women, although Foucault has argued that such 'power is not evil. Power is games of strategy [...] let us take sexual or amorous relationships: to wield power over the other in a sort of open-ended strategic game where the situation may be reversed is not evil; it's a part of love, of passion and sexual pleasure' (Foucault 1994 298). This notion of productive power gives rise to the subjects over whom, and through whom, power structures enmesh us all – both in fictional and in actual worlds.

It has been remarked that some feminist critics have exaggerated the role of feminism in SF, thereby excessively slandering the male of the species. For example, Anne K. Mellor has portrayed Frankenstein as the archetypal:

> [s]cientist who analyzes, manipulates, and attempts to control nature unconsciously engages in a form of oppressive sexual politics. Construing nature as the female Other, he attempts to make nature serve his own ends, to gratify his own desires for

power, wealth, and reputation (Mellor 1989 112).

This positioning superficially ignores the existence of female scientists who have objectives, desires, goals and ambitions identical to male scientists.

It is more reasonable to state that SF allows us to perform thought experiments that create altogether different utopias, as conventional utopias are often similar to Moore's *Utopia*, 'where equality is emphasized above all else, even to the point of suppression of individual liberty and imposition of a potentially oppressive conformity, […], and despite his imagination, Moore's Utopia is still a strongly patriarchal society' (Booker 1994 337-338).

In these ways, SF forces us into a deliberate consideration of where our actions, through a diversity of choices, might lead us, and perhaps guide us toward decisions that yield the greatest good for the many without pitting one gender against the other.

References:

Aristotle. *Poetics*. Else F.E. (transl.). Michigan: University of Michigan Press, 1967, 1994.

Asimov, Issac. "Satisfaction Guaranteed." *Amazing Stories*. April, 1951.

Atwood, Margaret. *The Handmaid's Tale*. London: Jonathan Cape, 1986.

Atwood, Margaret. *In Other Worlds: SF and the Human Imagination*. London: Virago, 2001.

Banks, Iain M. *Excession* .New York: Bantam, 1996.

Booker, Keith M. "Woman on the Edge of a Genre: The Feminist Dystopias of Marge Piercy." *Science Fiction Studies*. 21. 1994: 337-350.

Barker, Clive. *Weaveworld*. New York: Simon & Schuster, 1987.

Barr, Marleen S. *Future Females: The Next Generation* Lanham: Rowman and Littlefield, 2000.

Bujold, Lois McMaster. *A Civil Campaign*. New York: Bacn Books, 1999.

Corea, Gena. *The Mother Machine: Reproductive Technologies from Artificial Insemination to Artificial Wombs*. New York: Harper and Row, 1985.

Eagleton, Mary. *Feminist Literary Criticism*. Harlow: Longman, 1996.

Foucault, Michel. *Power/Knowledge – Selected Interviews and Other Writings 1972-1977*, Gordon C. (ed), Gordon C (transl.) and others. New York: Pantheon, 1980.

Foucault, Michel. *Ethics Subjectivity and Truth: Essential Works of Foucault 1954-1984.* Vol. 1, Rabinow P. (ed). New York: The New Press, 1994.

Haraway, Donna J. *Simians, Cyborgs, and Women: The Reinvention of Nature.* London: Free Association Books, 1991.

Freedman, Carl. "Science Fiction and the Triumph of Feminism." Science Fiction Studies. 81(27) Part2. July, 2000.

Hard, Robin. *The Routledge Handbook of Greek Mythology.* London: Routledge, 2004.

Herbert, Frank. *Dune.* New York: Ace Books, 1965.

Hussain, Rokeya Sakhawat. "The Sultana's Dream." *The Indian Ladies Magazine of Madras*, Madras, 1905.

Lefanu, Sarah. "Introduction" in Joanna Russ. *To Write Like a Woman: Essays in Feminism and Science Fiction.* Bloomington: Indiana University Press, 1995:vii-xii.

Le Guin, Ursula K. *The Left Hand of Darkness.* 40th Anniversary Edition. London: Orbit, 1969, 2009.

Lessing, Doris. *The Cleft.* New York: Harper Collins, 2007.

McIntyre, Vonda. *Dreamsnake.* New York: Houghton Mifflin, 1978.

McIntyre, Vonda. *Superluminal.* New York: Pocket Books, 1984.

Mellor, Anne K. *Mary Shelley: Her Life, Her Fiction, Her Monsters.* New York: Routledge, 1989.

Moore, Thomas. *Utopia.* Leuven, 1516.

Morris, Pam. *Literature and Feminism.* Oxford: Blackwell, 1993.

Pearson, Wendy Gay; and others. (eds). *Queer Universes: Sexualities in Science Fiction*. Liverpool: Liverpool University Press, 2010.

Robin, Peggy. *How to Be a Successful Fertility Patient: Your Guide to Getting the Best Possible Medical Help to Have a Baby*. Scranton: Harper Collins, 1993.

Piercy, Marge. *He, She, and It*. New York: Fawcett Crest, 1991.

Russ, Joanna. "When It Changed." in *Again, Dangerous Visions Book 2*. Ellison H. (ed). New York: Doubleday, 1972.

Russ, Joanna. *The Female Man*. New York: Bantam, 1975.

Russ, Joanna. "*Amor Vincit Foeminam*: The Battle of the Sexes in Science Fiction", in *To Write Like a Woman*: *Essays in Feminism and Science Fiction*. Bloomington: Indiana University Press, 1995: 41-59.

Smith, E. E. "Triplanetary." *Amazing Stories*. January-April: 1934.

Smith, E. E. "Children of the Lens." *Astounding Science Fiction*. November 1947-February, 1948.

Tiptree, James Jr. "Houston, Houston, Do You Read", in *Aurora: Beyond Equality*, McIntyre V.N. and Anderson S.J. (eds.). New York: Fawcett Crest, 1976.

Chapter 6. Legal Theory and Science Fiction: Law in the Eyes of Sci-Fi – Stefan N. Vella

Defining Science Fiction

Science Fiction is undefinable. Damon Knight in his book *In Search of Wonder* stated that 'Science fiction ... means what we point to when we say it.' He however, considers Science Fiction (SF) to be a field of literature worth taking seriously, and that ordinary critical standards can be meaningfully applied to it: for example, originality, sincerity, style, construction, logic, coherence, sanity, garden-variety grammar. He also claimed that a bad book hurts science fiction more than ten bad notices. Knight in fact stated that the term science fiction is a misnomer and he concluded that there were better tags devised, such as the term 'speculative fiction'. To this effect he stated:

science fiction is a misnomer, that trying to get two enthusiasts to agree on a definition of it leads only to bloody knuckles; that better labels have been devised (Heinlein's suggestion; "speculative fiction", is the best, I think), but that we're stuck with this one; and that it will do us no particular harm if we remember that […]

James. E. Gunn proposed an interesting definition of science fiction but he himself, again, echoed Knight's concern, that, amongst experts of SF, the task of defining SF is the catalyst of discord rather than agreement. He claimed:

The most important, and most divisive, issue in science fiction is definition....My involvement with definition may have begun with my original discovery of science fiction and my realization that this literature was different from every other kind...fantasy and science fiction belong to the same general category of fiction - that is, the fictional world represented is not the world of the here and now or even the there and then but the fantastic world of unfamiliar events or developments. (Gunn 2005:5)
In fact, Gunn, in his work *The Road to Science Fiction* refers to his definition , stating that:

Science fiction is the branch of literature that deals with the effects of change on people in the real world as it can be projected into the past, the future, or to distant places. It often concerns itself with scientific or technological change, and it usually involves matters whose importance is greater than the individual or the community; often civilization or the race itself is in danger.

In concord with was stated by the previous egregious exponents of SF quoted above, Everett.K.Blieler said that:

Science fiction is not a unitary genre of form, hence cannot be encompassed in a single definition. It is an assemblage of genres and subgenres that are not intrinsically closely related, but are generally accepted as an area of publication by a marketplace. Science fiction is thus only a commercial term.

Defining Legal theory

Legal theory is concerned with the functioning of the 'coercive normative institutions' in a given society (Hanoch and Roy 2011:672-691), dealing primarily with the concept of law and coercion, or law and force. The term 'legal theory' encompasses Philosophy of Law and jurisprudence, and is legal theorizing in its purest and simplest form (Solum).

Legal theory rests on two doctrinal pillars; the Positivist school and the Natural Law and Natural Rights school of thought. Two of the most known theorist adherents to the positivist theory of law are Austin and Kelsen. Austin approached the theory of law analytically. He pushed the exercise of analyzing law away from academic disciplines such as history, sociology, or secondary arguments emanating from political philosophy and calibrated his focus on key concepts such as 'law', 'right' , 'duty', and 'legal validity'. He shifted the focus of the understanding of law from ' a community-oriented approach' vision of law to that which is 'imperium oriented', thus viewing law as mostly the rules imposed from certain

122

authorized sources. Moreover, he changed the way of how law should be studied and analyzed.

This meant that instead of entering into the merits of analyzing law as a phenomenon, with a typical frame of mind of a moral theorist, or a political philosopher committed to the answering of questions tied to existential argumentation, Austin looked in the study of the nature of law as a scientific exercise. He analyzed law systematically. Austin adopted 'a command theory of law' (Austin). He considered law to be 'commands of a sovereign'. In this case 'commands' were defined by Austin to be an expressed wish for something to be carried out, combined with the willingness and ability to impose an evil if that wish is not complied with. The sovereign is a person who receives habitual obedience from the majority of the population, without habitually obeying any other person or institution. Austin's doctrine is epitomised in a proverbial dictum: "Law is commands joined to threats of punishment." Austin can be very much attributed to the attempt of defining legal theory as the science which defines law as it is, rather than as it should be.

Kelsen , in his bid to define law and legal theory, commenced by defining the 'object matter' of legal theory. His quote is clear and unequivocal: "A theory of law must begin by defining its object matter". Kelsen arrives at two features of law which are refutable with other social phenomena and are significant. He opined that laws are systems of norms for human behaviour and are coercive in nature. However, he adds to this definition the requirement of effectiveness, essentially, implying that there cannot be a system of norms without a basic norm with reference to which the validity of the other norms is determined and which itself is simply 'presupposed', together with his claim that an ineffective norm cannot be presupposed. Kelsen drew heavily from Kantian metaphysics. However, what must be borne in mind when analysing Kelsen, is precisely the fact that the institutions, practices and mores of society who duly proclaimed the law, are excluded from his conceptualisation of law. This means that he always failed to include in his analysis the reason why a particular law has been put into force or otherwise. What is also interesting to note is that Kelsen does not in any way enter into the merits of describing the consequences of the entry into force of a particular law (Kelson).

The Natural Law and Natural Rights school of thought, the second pillar in the analysis of legal theory, contemplates on the inherent aspects of the well-being and fulfillment of human beings and the communties they constitute. Natural law theories endeavour to establish principles of right action, specifying the first and most general principle of morality, namely, the obligation of one to choose and act in a manner that is compatible with a will towards integral human fulfillment.

A corollary of this application of the 'general principle of morality' is precisely the fact that there are human rights if there are principles of practical reason directing us to act or abstain from acting in certain ways out of respect for the well-being and the dignity of persons whose legitimate interests may be affected by what we do. Consequently, the natural law understanding of human rights is closely connected with a particular understanding of human dignity. This perception of human rights reveals the recognition of the natural human capacities for reason and freedom, being fundamental to the dignity of human beings, the same dignity which is protected by human rights. The basic goods of human nature are the goods of a rational creature. A rational creature is a creature who, if not impaired or prevented from doing so, naturally develops and exercises capacities for deliberation, judgment and choice (George 2007: 174-177). Naturally, the crux of the problem remains in indentifying the source of all norms we consider constituent of natural law. When concerned with natural law there are philosophers who are theists and those who are not theists. The former believe that natural law emanates from Divine law and therefore Divine law is the source of natural law. The latter believe that one must use reason to discover the laws governing natural events and applies them to thinking about human action . One can infer the existence of a corpus of moral norms, including norms of justice and human rights, that can be known by rational inquiry, understanding and judgment even without any special revelation, and which can also provide the basis for an international regime of human rights (George 2007: 82).

The limitations of Legal theory

Legal theory explains law and its implementation. In its quest of being uncontaminated by other disciplines, and to be narrowly focused and presumably, without undue distractions on its definition of the concept of law and its inherent characteristics, it failed miserably to enter into the question of defining legitimacy as witnessed through the persistent error of having first to define law before assessing its role with the rest of the current social phenomena.

Fuller defines law as being the 'enterprise of subjecting human conduct to the governance of rules'(Fuller 1977: 96). Fuller reiterates that the legitimate purpose of developing jurisprudence is to 'analyze the fundamental problems that must be solved in creating and administering a system of legal rules.' In so doing he identifies in his doctrine, a purpose in law which is the organizing principle and which he defines as the 'internal morality of law.' Fuller identified the sole purpose of law as being 'a modest and sober one,'one which subjects human conduct to the guidance and control of general rules"(Fuller 1977:146).

Finnis on the other hand tied the definition of law to the concept of the latter governing a complete community, therefore stating:

That is why it is characteristic of legal systems that (i) they claim authority to regulate all forms of human behaviour (a claim which in the hands of the lawyer becomes the artifical postulate that legal systems are gapless); (ii) they therefore claim to be the supreme authority for their respective community, and to regulate the conditions under which the members of that community can participate in any other normative system of association; (iii) they characteristically purport to 'adopt' rules and normative arrangements (e.g. contracts) from other associations within and without the complete community, thereby 'giving them legal force' for that community; they thus maintain the notion of completeness and supremacy without pretending to be either the only association to which their members may reasonably belong or the only complete community with whom their members may have dealings, and without striving to forsee and provide substantively for every activity and arrangement in which their members may wish to engage (Finnis 1992:148-149)

This quote echoes the Weberian analysis of law, so much so, that Weber an exponent in the field of sociology, proposed his 'Theory of Man' outlined in his four characterisations of human action. The first category of actions is the 'goal-rational actions'. This mode of action entails the choice of the most effective means to reaching one's goals, being logical, scientific, and economical. The second category of actions is the value-rational mode of action. In this mode of action, the most effective means is chosen, so that a morally good objective must be attained only by a morally good means. The third category of actions is that category of actions which are mainly concerned with emotions. Here conduct is only emotional and hence non-rational. Finally, the fourth category of actions is the 'traditionalist' mode of action. This covers habitual conduct which can be correlated with the exitence of specific practices and respect for existing authority. These four categories of actions are means of how human beings make sense of their actions. Weber believed that it is fundamental to man's nature that he seeks to give some sort of sense to his life (Campbell 1981: 176-178).

It is to be noted that Weber, who was highly influenced by Nietszche, did not believe in any universal set of values which humans are bound to adopt. Naturally the negation of any set of universal values implies that there is no single definition of good and evil. In fact, Nietzche, is quoted as saying: O my brothers is everything not more in flux? Have not all railings and gangways fallen into the water and come to nothing? Who can still cling to "good" and "evil".. There is an old delusion that is called good and evil. Up to now this delusion has orbited about prophets and astrologers. Once people believed in prophets and astrologers and therefore people believed "everything is faith, you shall for you must!". Then again people mistrusted all prophets and astrologers and therefore people believed "everything is freedom you can for you will!.

O my brothers, up to now there has been only supposition not knowledge, concerning the stars and the future, and therefore has hitherto been only supposition, not knowledge concerning good and evil!"(Nietzche 1971:219)

Weber, in his Theory of Society, which is a direct compliment of his Theory of Man, defines a 'social relationship' as the 'behaviour of a plurality of actors in so far as, in its meaningful content, the action of each takes account of that of the others and is oriented in these terms'. 'Social relationships' are of three forms: 'conflict,' 'community,' or 'association.' This is deemed as the backbone of the analysis on the coercive element of law. According to Weber conflict 'is oriented intentionally to carrying out the actor's own will against the resistance of the other party or parties'. To do this successfully is to exercise 'power' and so achieve 'domination' or 'imperative control'. On the other hand, a social relationship is 'communal', if its direction 'is based on a subjective feeling of the parties, whether affectual or traditional, that they belong together'. Last but not least, a social relationship is 'associative' when 'the orientation of social action within it rests on a rationality motivated adjustment of interests or similarly motivated agreement.' The crux of Weber's Theory of Society is the idea of valid norms or legitimate order. Weber surmises that authority exists when the social actor behaves predictably because for some reason he believes in the legitmacy of certain rules or practice. Physical power has social significance when it is believed to be legitimate, in the sense that, it is exercised in the context of the political authority of the state with its monopoly of coercive force in a given territory. Weber outlines three ideal types of legitimate order or authority; charismatic authority, traditional authority and legal authority. In rational or legal order it is possible to know which rules are 'formally correct and have been imposed by an accepted procedure'(Cambell 1989:181). In a legal authority the ruler has to obey the law if he is not to lose the capacity to be obeyed. Max Weber's analysis is pertinent to understand how law features in a socio-legal perspective. The problem with Weber's analysis is that it is more focused on the influence of law in economic development rather than the study of law itself. Moreover, Weber fails to give a complete definition of law, in no instance does he define law, but he addresses the archetypal role of authority in the exercise of enforcing law. This means that Weber defines the most evident feature of law. He enphasised as did other theorists in the branch of legal theory, the coercive function of law. His conception of law is epitomised in the quotation ' an order will be called law if it is externally guaranteed by the probability that physical or psychological coercion will be applied by a staff of people in order to bring about compliance or avenge violation'(Weber 1922: 32).

It is evident that 'coercion' is only one element of the anatomy of law, and this was clearly understood by Weber. Weber, in actual fact, considered law to be one form of 'legitimate order,' a term Weber uses to refer to any structured source of guidelines for right conduct. This means that in the Weberian scheme, law is a subclass of a category called legitimate or normative orders. In fact, the Weberian scheme can be explained as follows:

Thus, in the Weberian scheme, law is a subclass of a category called legitimate or normative orders. All such orders are socially structured systems which contain bodies of normative propositions that to some degree are subjectively accepted by members of a social group as binding for their own sake, without regard for purely utilitarian calculations of the probability of coercion. 'Law' is distinguished from other forms of normative orders on the grounds that it additionally involves specialized agencies enforcing norms through coercive sanctions. 'Law' ...is simply an 'order'," he [Weber] said, 'endowed with certain specific guarantees of the probability of its empirical validity (Max 1972: 726).

This means that Weber understood law to be, like custom and convention, one of the basic sources of normative guidance in society 'a place where men look to determine how they ought to behave.'

'Orders' which have coercive powers were called law but not all law is coercion (Max 1972: 726). Weber in actual fact, defines law as one form of legitimate order which carries the attire of coercion but has the function to guide human beings within a social order how they ought to act. Thus law is the measure of what is the standard of good standing out from other norms because its non-observance will be chastised with some form of punishment (coercion). However, his analysis was vital because it drew out clearly the fact, that law is not only coercive in nature, but it has the didactic function, of the 'acceptable norm,' or what one simplistically considers it to be 'basic morality.'

Dennis Lloyd in his book *The Idea of Law,* discussed in depth Max Weber's action theory. He criticised Weber's theory as being too neat, and apparently too perfect for it to fit in the many historical situations of true and active societies. Weber fails to explain the link between legitimacy and coercion. Lloyd was pertinent in pointing out that coercion or 'force' does not necessarily prove the existence of legitimacy. This means that who exercises force does not necessarily have power emanating from legitimacy. Law imposed by a dictator or a tyrant will be obeyed by his subjects, without necessarily believing in the legitimacy of their ruler. Lloyd further claims that there is a justified objection to equate law exclusively to coercion. He divides the exponents of this view into two distinct categories - the moralists those adhering to the view that there exists a social contract through which man, being free in spirit agree to submit to law and government. The moralists believe that law cannot be based on force because force in itself is wrong. To support the argument of these moralists, is the principle that no system of rules can qualify as law unless it coincides with, or can at least be subsumed under the rule of morality. Yet, this line of thought presents a series of problems. First there is the relationship of law with morality. Those theorists attributed to the school of thought promoting the concept of the existence of a form of social contract, reiterate that people obey the law not because they are constrained to do so by force but because they consent or at least acquiesce to the operation of law, and it is this consent rather than any threat of force which causes the legal system to work. Lloyd contends:

The question remains as to what justification there may be at the present day for insisting on the inclusion of the element of coercion in our model of the law. In the international sphere coercion, as we have seen, plays a small part, and even in national law it is generally recognized that people usually obey the law because it is the law, and not just because they are afraid of being punished if they disobey (Dennis 1964: 40-41).

A reading of the various theories of the concept of law demonstrates the consistent wrong doing of paying little attention to the distinction of four fundamental concepts: validity, normativity, content and legitimacy. Ironically these are extremely important components of legal theory itself. The lack of distinction between these concepts brought about a nebulous description of law and its function, little being said on the characteristics of a perfectly legitimate sovereign or law maker, or rather how the source of law should be for the derivant norm to be considered 'valid' and 'legitimate.' The content of a legal norm is that which the norm prohibits, normally by attributing certain sets of facts that have to be obtained for certain legal outcomes. The normativity of a legal norm is the spirit in which the will of the law is expressed in such a manner as being a matter of 'it ought to be done' rather than 'it may be done.' The question of validity is a matter exclusivley concerned with the fact that an order is law because it is part of a legal system, otherwise it is not law at all. Finally, legitimacy is concerned with the question of when the issuer of legal norms is entitled to make such demands.

An accurate account of what is law was given by Dworkin in his work *Law's Empire* which was an attempt at building a solid analysis of the concept of law on the above mentioned concepts. Dworkin's analysis was important because it explained eloquently the relationship between legitimacy and law. Dworkin manifested clearly that jurisprudence (as part of legal theory) is concerned with the explanation of the relationship between the validity, normativity, legitimacy, and content of legal norms. Dworkin stresses that even if convention gives comfort to and offers a satisfying answer to the question of the normativity of law, the existence of such a convention matters only if one can provide the criteria which a particular norm reaches for it to be considered legitimate. The question of legitimacy can be raised with regard to every legal norm and it explains Dworkin's claim that jurisprudence is properly understood as a branch of political morality. Particularly, it explains why the question of legitimacy can be asked (and in practice is frequently asked) both at the level of entire legal systems and at the level of particular cases and it explains the nexus between interpretation and legitimacy.

Dworkin consistently shunned the identification of law as a simple fact, but stressed on law as being in itself, a statement of some form of values. He therefore considered law to be analysed properly once it is analysed within a context of an all-encompassing moral theory. The total drift from the definition of law is seen as being essentially 'coercive' in nature, but law is also seen as the reflection of identifiable moral principles. This is a view which converges with Fuller and to a certain extent Weber. Dworkin looks at law from a 'normative' point of view, and cogently, justifies the subjection of law to an interpretive exercise. To this effect he states that law is constituted of rules and principles. These principles are moral principles which confer rights on individuals. According to Dworkin, the judge does not make law, but he reveals the law, he exposes what the law is, by precisely, interpreting the law in the light of the moral principles which are incorporated in the law itself.

Dworkin considered the conception of law as being interpretative in nature, just like any other political conception, which has its fundamental propositions profoundly disputed. Dworkin thus contends that law is put into action, through a reconstruction employed by the interpretative activity. To this effect he said:

Law is an interpretative concept. Judges should decide what the law is by interpreting the practice of other judges deciding what the law is. General theories of law for us are general interpretations of our judicial practice...We noticed how convictions about fit contest and constrain judgments of substance, and how convictions about fairness and justice and procedural due process contest with one another. The interpretative judgment must notice and take account of these several dimensions...But it must mould these dimensions together into an overall opinion: about which interpretation, all things considered makes the community's legal record the best it can be from the point of view of public morality"(Dworkin 1986: 411).

This implies that Dworkin deals with the definition of the nature of law as a doctrinal construction that is internal to the legal discourse, but can also be seen from a strictly theoretical point of view. So, the 'normative concept of law,' is sufficient for a defined characterization of law. If law is seen as a system of rules, the correctness of a norm is established when put in juxtapposition with a comprehensive theory that gurantees the same raison d'etre of the law's existence. Dworkin considers a theory of law to be a theory of public morality (Caballero et al). Dworkin's theory despite its many strong points lacks considerably in its exposition of legitimacy. Dworkin attributes legitimacy to social practices, and this thesis lacks in its strength and is extremely dangerous to be translated in practice. This is particularly so, if so called 'public morality' is equated to 'social practices.' If legitimacy is tied to social practice, in its strict and literal sense, legitimacy has to be admitted to many situations where minorities are severly discriminated without any sound justification.

Therefore, the concept of legitimacy remains orphan once again in the field of legal theory.

SF with its own fixed paradigms touches on two fundamental issues - the sovereignty of law and the definition of unjust law.

Law as depicted in SF
Law as the Guiding Light and the Gurantee of Order in SF Utopias
SF envisages two forms of utopia; the traditional utopia which envisioned ways in which human society might be re-organised on earth those utopias which take a more modern form. The latter form of

utopia envisages a society which can be perfect as a result of evolution and technology. The traditional utopia included law as the means by which order and proper regulation is in place.

Law in this case is given the didactic role, the role of guidance, indicating to the individual right from wrong, and therefore is considered to be the essential component preserving the dignity of man. SF in the creation of traditional utopias implicitly acknowledges the fallibility of man.

There were various philosophers who shared the view that man is fallible. Hobbes is quoted as saying: '[…] No arts no letters, no society and which is worst of all, continual fear and danger of violent death and the life of man, solitary, poor, nasty, brutish, and short' (Hobbes 1651: 110). Paul Ricoeur also opines "What is meant by calling man fallible? Essentially this: that the possibility of moral evil is inherent in man's constitution.... (Ricoeur 1986: 133). In his argument that man is finite and fallible, St Augustine claims that man is not the measure of all things, adding further that man is characterized by a nature which is evil (Chadwick).

To this effect, SF deals with the concept of the peril of having fallible man being the creator. SF deals with the function of law as being the means with which the human race is brought back in line. Its transgression blurs the thin line of delineation between man and animal. Law in the eyes of SF is the instrument with which monsters or monstrosities created by man are put in check. The basic understanding of law as portrayed in SF, is that through law, man is shown the minimum standards of morality, 'the basic morality'. Once man fails to adhere to the basic standards of morality, man loses his stature and reverts back to his animal state. This is cogently explained in the novel by H.G.Wells *The Island of Dr.Moreau.* Dr.Moreau creates a number of human-animal hybrids. In his island, the human-animal hybrids learn through law how to live in society. Eventually, transgressions of the law lead the hybrids to reject their human status and return to their animal natures (Travis).

A traditional utopia designs mechanisms which involve legislation, education or institutional changes and occasionally changes technology or environmental management. The traditional form of Utopia is represented par excellence in Thomas More's Utopia. In More's Utopia the strict laws are vital to restrain 'original sin'. He projects the possibility of each member of this 'community' working for the common good and law is enforced by the same members of the community through vigilance against signs of disobedience for which law is clear to all and sundry, a principle envisaged in the principles of the Rule of Law doctrine. A more recent example of the above mentioned category of utopia is *Blue Mars*. In *Blue Mars* the process to create a new society which will avoid the failures of Earth societies was identified. It contains a description of the creation of a utopian constitution for the newly terraformed planet. It does not only include a set of political arrangements, but also a list of human rights and human obligations, and a list of the rights of the Martian landscape. This implies the belief in SF that law is there to instill order and to create obligations which one is expected to follow to guarantee the well-being of the other. This undoubtedly is a natural law conception of law (Edward).

The relationship of Man and Technology in SF: The Frankenstien Myth

The question of legitimacy of law in SF is tied to the concept of urgency to act. In SF the call for the need to intervene is usually explained in the abusive relationship of man with technology. This abusive relationship of man with technology, was defined in Heidegger's work *The Question Concerning Technology*. Heidegger stated that technology is a human tool and a means to an end. He considered the concept of technology to be an anthropological term denoting a network of tools and equipment at the disposal of mankind and is a human activity in the broadest sense of the word.

Unfortunately, according to Heidegger this conception of the notion of technology feeds us the wrong assumption that technology is something we control, something we can master and bring under our way as it facilitates our efforts to secure certain objectives.

Technology must be seen in a different light; it is a way of revealing the world we live in, and therefore, the essence of technology is the realm of truth. The problem is how modern technology reveals itself to the world. According to Heidegger modern technology does not stop at being provided with what nature produces but it looks into the means of manipulating nature, to impose upon it, to undermine its ontological and structural integrity, so that modern man can demand more of it and set upon it with the relentless zeal of a 'grand inquisitor'- an interesting term coined by Heidegger - where the scientists are the inquisitors and also the henchmen, vivisecting every corner of the earth's structural composition for no other reason. Heidegger was worried with the abuse of technology by man where technology was being exploited to its maximum not for the purposes of 'bringing-forth' but as mere production (Mahon).

The Frankenstein myth pushes us into a stark reality in the guise of a biblical admonition. On the one hand technology is projected as a monstrosity, and the controlling law a monster in itself. Mary Shelley's *Frankenstein* contains very well defined elments which are widely known and undisputed, and follow a strict sequence: A scientist creates a monster, the scientist entices the monster his creation, the monster learns about humanity and faces its own monstrousness, the monster develops a pathology and it ends up killing its creator. Victor Frankenstein epitomises the modern scientist who is more concerned with technical concerns rather than the consequences of his creation. The ambivalence in the creation is highlighted; it can be good, but at the same time it can be evil. On the one hand the monster which is the reifecation of technology, is something that is to be loved and at the same time feared. On the other hand, the creator is the monster. He represents the egoistic nature of man who created something, which cuts itself off from humanity and wants a mate of its own, therefore, explaining the amoral and dehumanised conceptualisation of technology. This signifies the possibility of having technology overcoming man, and therefore being vulnerable to the bloody product of science itself. In the Frankenstein myth, law is the saviour but is in itself a 'monster' in disguise. It is neither good nor evil, but pure power, which through the lawyer/technocrat turns out to be beneficial to the passive society, a society which is numb in the face of technology's dominion. Law is seen as power which can be used to achieve good within society.

Law is projected as being the shield of society against the depredations of a monstrous yet benign technology (Kieran 2007).

The Relationship between Law and Technology : The Heidegerrian Legacy

The relationship between law and technology is explained by three schools of thought: one which considers law as technology; a post-Heidegerrian theory which envisages absolute technology and enslaved humanity; and a post-Heidegerrian theory which is constructed from the phenomenological analysis of technology, focused round the 'being in the world' concept. The concept of a technological utopia fits in the latter school of thought.

Law as Technology: The Dystopian definition of Sovreignity

The first strand is made manifest in the work by George Orwell *1984* . The novel revolves around a character Winston who lives in Oceania. It is modelled on the idea that one party controls a society, a nation, a territory; the classical elements of a State. This concept is understood in literature concerning legal theory as being a legal system and a rational system of rules. The precepts that constitute the root of this concept are represented with coercion. Drawing from Weberian analysis, the presence of a centralized organized force fulfill the rules that harmonize the system in all its parts, giving to the community social peace. Thus the State is understood as orientated to the 'common good.' This explains Kelsen's doctrine of real power. Kelsen refers to a tripartite perspective, the State with this power validates the legal system, constituting a natural force, resulting in irresistible dominion. In Oceana, the elite in rule create a language of their sort so as to eliminate any vocabulary which in some way or another arouse the will for rebellion. Subversive thoughts was considered a crime. Expressions of love were also a criminal act. It was a society with a capillary system of surveillance.

Winston, initially rebelled, yet he was overpowered by the system, accepting it and considering it as morally just. In Oceania, the law claims authority over the community and the life of the individual, his morality, his language, his individual expression and his actual way of thinking, such that the individual is left with no choice but to accept the system the way it is. Raz contends that the law claims authority (Raz). It has to be comprehensive in that all aspects of social life need to be regulated, whether by prohibition, requirement or permission. The inherent feature is force, rendering law to be supreme in a legal system sitting at the top of the hierarchy of the social structure. The concept of obedience and what means to follow a rule depends on sophisticated machinery established by the institution, and beyond, by the leader as the 'man behind the curtain,' in the

Wizard of Oz analogy. Oceania despite its evil legal system created its own 'virtue' upon which all of its rules are legitimized and followed by the masses. The system and its subjects think that the behaviour of the 'Big Brother' is legitimate and acceptable, and the dissent of an individual will seem to be arbitrary juxtaposing the system in a way that is considered correct. So, the law is seen to be within the prescribed limits because it follows the same logic of the Rule of Law, in that the latter is conventional, objective and delimited. Government action must have the foundation in law and must be authorized by the law. The people should be ruled by the law and obey it, and the law should be such that the people will be guided by it. The law must enable obedience.

Technology is Law: The Dystopian definition of Unjust Law

The second strand is manifested in the list of films to be expounded hereunder which outline the irrelevance of law when the letter is a tool of attrition, that is, when the law is unjust, a law which will prescribe the slavery of mankind. In the natural law and natural rights doctrine, unjust law is no law in the first place (Finnis). In these films technology is presented as something evil that escapes from the control of its creator (the human being) and is used to exploit man rather than serve him. Inevitably, the conception of Law as a just order collapses in the face of technological domination and instead the Law becomes a tool in the service of the power that is conferred to those who control it. On more than one occasion, the entity controlling the technology is the technology itself. In this perspective, technology is a sign of unjust Law.

In James Cameron's *Terminator* a computer (Skynet) dominates the world using robotic warriors to ensure that humanity remains subjugated to the reign of technology and the machines it creates. In *The Matrix*, from the Wachowski brothers, computers converted humans into 'batteries' to supply energy forcing them to live in a continuous computer-generated dream state, a virtual reality, typical of a computer game. In Fritz Lang's *Metropolis*, a minority who controls the alienated masses attempts to maintain their power by creating a robot to replace Maria, a charismatic leader who represents a threat that could lead to the liberation of the oppressed. In George Lucas' *Star Wars,* a dictatorial empire, under the direction of a being that is half man, half robot (Darth Vader), subjugates entire planets with technological firepower never imagined before. In Steven Spielberg's Minority Report, technology, coupled with the mental power of human mediums, makes it possible to convict and incarcerate suspected delinquents before they commit any crime and without granting them the right to defend themselves. In this case we find that technology is a mechanism that concentrates power, the few whether dictatorial governments or enormous, unscrupulous corporations are the ones who enjoy technology and employ it to reach their goals. Technology is also used to threaten or limit individual rights. The inevitable result of this concentration of power is disempowering individuals. Further the political structures presented reveal a malevolent coalition between powerful political and economic figures who join together to exploit the weak, legal institutions, such as property and contracts, and are characterized such that they become part of the unjust mechanisms of oppression. Finally the 'good guys' are portrayed as people who lack power and operate with very few technological resources and are forced to become outlaws, to subvert the established order (Bullard).

The Code within the Law : The Three Laws of Isaac Asimov

The Code – The Three Laws and the Robot Series

The third strand focuses on Isaac Asimov. Asimov, was largely known for the Three Laws and his Robot Series. These three reputable laws which are worth mentioning here are: a robot may not injure a human being or, through inaction, allow a human being to be harmed; a robot must obey the orders given to it by human beings, except where such orders would conflict with the First Law; and a robot must protect its own existence as long as such protection does not conflict with the First or Second Law.

Asimov, cunningly, employed their iterations as a kind of algorithm for the construction of possible robot stories, which stories re-affirmed these principles rubbing them deep in the formative clay for undisputed Laws. These laws evolved in a form of Code which, irrespective of their application in real life, remained clearly understood, justified and taken as read. This Code featured in the early works of Asimov specifically *I, Robot* and *The Rest of the Robots*, and the novels *The Caves of Steel* and *The Naked Sun*, presenting an image or vision, of the artificial being succumbing to the 'iron law of the Laws' in order to 'remain helpful servants of humanity' (Clute). The Robot series were a manifestation that 'good' robots can exist, and robots themselves are written in a fashion that claims sympathy rather than fear. Therefore, by the time of the great robot novels of the 1950s, this *ratio legis* of the Code was ultimately crystallized in the figures of Earth detective Elijah Bailey and his partner R.Daneel Olivaw. Here the reader of SF was groomed to accept the image of the docile robot. Eventually, robots in the later novels of Asimov, are no longer considered to be servants, but are ascended to the order of keepers, in that in complete obedience of the Code, will need to protect interstellar humanity from itself (Clute).

The Code within the Law – The Robot and the relationship with Man

Asimov built a parallel legal system (which is explained further on) applicable to human beings which follows the same principles enunciated by Raz and Kelsen. The difference which is fundamental is that these Laws are applied by robots to be enforced on robots. Yet they do have a bearing on law as applied to human activity. The next obvious question becomes then on how a human being is defined. As part of a natural component of his doctrine of soveriegnty of Law, Kelsen had to define a human being in juxtaposition to the concept of personhood. In fact he states that:

To define the physical (natural) person as a human being is incorrect, because man and person are not only two different concepts but also the results of two entirely different kinds of consideration. Man is a concept of biology and physiology, in short, of the natural sciences. Person is a concept of jurisprudence, of the analysis of legal norms (Kelsen 1945: 93-95).

This means that the human being, as a biological entity is a different entity than the physical person in legal terms, with the human being becoming the basis of the physical person in legal terms as a symbolic and linguistic unity. The biological human being is only the enclosing line of a physical person in legal terms. The human being exists for the law only for the limited extent to which rights and duties refer to him and the physical person in legal terms and the juristic person are both legal creations having in common the character of artificiality. This signifies an immersion in the big conundrum of philosophy and cybernetics and in a limited sense robotics. A physical person is not and cannot be a human being and a natural entity. It is the construction of legal theory in a specific historical situation and according to the knowledge revealed through science and technology at that particular moment in time. So, physical persons are 'legal fiction' and this means that anything not human can be a legal agent. Therefore, the question is under what conditions this is possible and what required features and abilities are to be demonstrated in this 'physical person' qua 'agent'. This enters into the remits of defining the nature of robots. Robots are not persons, but they are things. However, they are animate objects and therefore they can act meaning they can be agents. In this case, however, robots can be agents when they are instructed by man to do something. And here, we see how law and the Code of Asimov are at parallels. The law (here referring to the corpus of norms governing human conduct) is concerned with the robot in so far as it is considered to be the extension of man, something that is man's creation. The liability incurred here is on the creator. If the robot is operated in such a manner as to cause damage to others, due to lack of expertise, recklessness, or unethical use, the creator is held responsible. Asimov however constructed a different law; he speaks of a law which regulates robots internalized and observed by the robots themselves, implying that robots here are not only agents, but active subjects of a form of legal regimen. Therefore, the Asimov rules are vital in the legal field because they set out the pretext for the regulation by law of the consequences of the activity of the agent (therefore of the robot acting 'literally in the hands of the artificer' man himself).

Conclusion

When analysing how SF defines law, one is bound to understand clearly the relationship of legal theory with SF. SF contributes to the unequivocal explanation of what a legitimate law must not do, by not being unjust in its nature and implementation. SF sets out clearly the features of unjust law. Legal theory in this regard provides the bulwark upon which one constructs the proper understanding of a legitimate legal system and its function as explained in SF. SF, on the other hand attempts at defining legitimacy, which unfortunately, legal theory fails to do.

Bibliography

Austin, John. *Stanford Encyclopedia of Philosophy* 2014. Web, 28 February 2015.

Blieler, Everett Franklin. *"Science Fiction: The Early Years."* Kent: Kent State University Press, 1990.

Bullard, Alfredo, G. *Law in Science Fiction – Law and Technology: From Socialist Dystopia to Capitalist Utopia.*
<http://www.law.yale.edu/documents/pdf/sela/SELA11_Bullard_CV_Eng_20110325.pdf>

Caballero, Cecilia and Almeida, Lois Danilo. *Legal Theory and Epistemic Values: Against Authoritarian Interpretativism,.* 25th IVR World Congress LAW SCIENCE AND TECHNOLOGY, Frankfurt am Main 15-20 August 2011.

Cambell, Tom. *Seven Theories of Human Society.* New York: Oxford University Press, 1989.

Clute, John. *Isaac Asmiov. A Companion to Science Fiction.* New Jersey: Blackwell Publishing, 2005.

Dagan, Hanoch and Krietner Roy. "The Character of Legal Theory." *The Future of Legal Theory* 96: 19 (2011):671-9.

Dworkin, Ronald. *Law's Empire.* Harvard: Belknap Press of Harvard University Press, 1986.

Finnis, John. *Natural Law and Natural Rights.* New York: Oxford University Press, 1980.

Fuller,Lon.L. *The Morality of Law.* New Haven: Yale University Press, 1977.

George, Robert, P. "Natural Law." *Harvard Journal of Law and Public Policy* 31 (2007):174-177.

Gunn, James E. and Candalaria, Matthew (eds.) *Speculations on Speculation: Theories of*

Science Fiction. Maryland: Scarecrow Press Inc., 2005.

Hare, R.M; Barnes, John; Chadwick Henry. *Founders of Thought: Plato, Aristotle, Augustine*
New York: Oxford University Press, 1991.

Hobbes,Thomas. *Leviathan.*
<https://scholarsbank.uoregon.edu/xmlui/bitstream/handl e/1794/748/leviathan.pdf> (online), last visited 30 September 2015, 1651: III.110.

James, Edward. *Utopias and anti-utopias*; *The Cambridge Companion to Science Fiction*. Cambridge: Cambridge University Press, 2003.

Kelsen, Hans. *General Theory of Law and State*. Part One, Chapter IX, 1945.

Knight, Damon. *In Search of Wonder*. Chicago: Advent Publications, 1952.

Lloyd, Dennis. *The Idea of Law*. Middlesex: Penguin Publishers Ltd, 1991.

Micahel, Travis. "Making Space: Law and Science Fiction." *Law and Literature*. Cardoso School of Law of Yashiva University, 23:2 (2008): 246.

Minn.J.L. *Sci&Tech* 8:2 (2007): 449-474.

Nietzsche, Friedrich. *Thus Spoke Zarathustra* . Translated with an Introdcution by R.J.Hollingdale. London: Penguin Publishers Ltd., 1969.

O'Brian, Mahon. *"Commentary on Heidegger's The Question Concerning Technology"*, IWM Junior Visiting Fellows' Conferences, Vienna, XVI:1 (2004): 1-39.

Perez Lopez, Nicolas. *"Dystopias and Legal Theory: A view from Orwell's 1984."*Universidad Carlos III de Madrid.
http://www.academia.edu/5898064/Dystopias_and_Lega l_Theory_A_View_from_Orwells_1984> (online), last visited 30 September 2015: 1-24.

Raz, Joseph. *The Authority of Law*. Oxford: Oxford University Press, 2009.

Ricouer, Paul. *Fallible Man: Philosophy of the Will* . Translated by Charles.A. Kelbley. New York: New York Fordham University Press, 1986.

Solum, L. *Legal Theory, Legal Theory Lexicon*. (online) 2005. Web. 27 February 2015.

Tranter, Kieran. *Nomology, Ontology, and Phenomenology of Law and Technology*.
(online), last visited 30 September 2015: 1-35.

Trubek, David. M. *Max Weber on Law and the Rise of Capitalism*. Yale: Yale Law School, 1972.

Weber, Max. *Economy and Society*. California: University of California Press, 1922.

Chapter 7. 'Hail Columbia! Vision of a Great City'. A taxonomic appraisal of American cities of the imagination in visual science fiction – Stefan Rabitsch

Introduction

It is easy to *imagine* the American city. Impressive skylines that are populated by skyscrapers and towers built out of concrete, glass and steel, and interspersed with iconic landmarks, readily come to mind. An imagined flyover reveals the deep canyons that streets carve into the phallic geography of American economic power. While being neatly arranged on the seeming conformity of the grid, the city is bisected by arterial expressways. At the bottom of these canyons, a multitude of people go to and fro work and play, feeding off and contributing to a supposedly healthy consumerist metabolism. Yet, these images usually apply only to the downtown areas of American metropolises, especially to the throbbing centers of corporate and political power. The vertical density of downtown quickly gives way to the equally dense horizontalism of outlying neighborhoods and boroughs. Each lays claim to one or more distinguishing characteristics that are circumscribed by ethnic, gender, artistic and economic communities. They are constantly besieged by the forces of gentrification and (re)development. Moving out even further beyond a cordon of beltways and a connecting web of freeways, we find hubs of industrial activity, shopping malls, and ultimately the homogeneity of the suburban sprawl; it is there, the "good All-American life" is sold to those who can afford it. Despite a city's design and presumed order, when viewed from an orbital vantage point, the American city displays the organic qualities of cell clusters that reinforce its living "nature".

The relative ease with which these images are conjured speaks to the pervasiveness of the American city as it comes by way of popular culture artifacts rather than direct experience. After all, the United States is often perceived as one vast urban space that is recreated from images seen in movies, TV-series, music videos, video games, and a host of advertisements. In short, American cities appear to be everywhere whether one resides in the U.S. or not; from the flashy opening credits of the various incarnations of *CSI* and the ominous geography of Washington in the opening sequence of *House of Cards* to the desolate urban wastelands of Boston, Pittsburgh, Salt Lake City and Las Vegas in *The Last of Us* and the *Fallout* video game series, respectively. The latter two point to an implied tautology.

It is safe to say that reflections, refractions and re-imaginings of the American city provide the archetypal template for the bulk of urban spaces

157

found in the visual science fiction imaginarium. With Metropolis, Coruscant, Gotham City, Mega-City One, Panem's Capitol, Basin City, the Sprawl, the Grid, Caprica City, the Capital Wasteland, Nos Astra and the Citadel serving as a non-exhaustive list of examples, it is clear that visual SF—movies, TV-series and video games—teems with American urban environments. This has chiefly to do with the American city's inherent historical trajectory towards the future and a latent modern(ist) semblance of utopia that it has yet to shed. The American city of the visual SF imagination literalizes the metaphoric characteristics that have been ascribed to the American city and to America-as-a-city since the 17th century.

Less constrained by realist conventions, the SF city is able to "realize" more fully and thus literalize the *palimpsestuous* nature of the American city as the New Jerusalem that coexists with the New Sodom and Gomorrah.

Channeled through the historical "moment" of the modern American city at the end of the 19th century as the genesis of the American city of the visual SF imagination—the default starting point of any previous study of the city in SF—this essay aims to *rehistoricize* this quintessential *topos* in SF by following two as yet untapped historio-cultural trajectories: 1) The Puritan imagination and rhetoric, which are modified by 2) the antipathy towards urban environments which is paradigmatic of the Jeffersonian pastoral vision for America. When attempting to map the American city both in as well as outside SF, we find that there are a number of overlapping dualities that make it perhaps difficult to read at first: high vs. low, center vs. periphery, inner city vs. suburbs, vertical vs. horizontal, uptown vs. downtown, feminine vs. masculine, rich vs. poor, and dark vs. light.

While acknowledging the *dualism* inherent in American cities, both real and fictional, this essay aims to offer a more diversified taxonomy of SF cities. Instead of replicating the dualistic histories of the imagined city of light vs. the dark city that has become the default approach for this *topos* deriving from SF movie criticism, a concise taxonomic survey of four readily recognizable urban geographies in visual SF will be offered: 1) urban utopias, 2) urban wastelands which may or may not be postindustrial or post-apocalyptic, 3) sky cities and cities in outer space, and lastly 4) virtual(ized) cityscapes. Both the broadening of media boundaries to include more than just SF films and the diversification of taxonomic parameters are geared towards a renewed appraisal of the SF city which combines aesthetically derived categorization with culturally contextualized and historicized urban ontologies.

To engage in reading any city and especially the palimpsestuous character of the American city as it is literalized in visual SF, is fraught with challenges that makes necessary two disclaimers about the power of discourse and the nature of utopias. Roland Barthes claims that the city is a text in that it is 'an inscription of man [and woman] in space' (Campbell and Kean 188). When a geographic entity is labelled "the American city", meaning and coherence are assigned to it, and this is hardly intrinsic to its character. The discourses used are not about the American city per se, but they rather constitute it for readers/viewers/gamers/*interacteurs*.

Consequently, any readings of the city can only ever be temporary, like an Instagram snapshot. To this day, the American city retains a semblance of a modern(ist) utopia. Utopias and utopian thought present a minefield of scholarly criticism, and as such lie beyond the scope of this essay. In the vein of Linda Hutcheon, the key to most utopias is their inherent reactionary and nostalgic outlook. Most utopias, especially in SF, are not about what could be gained in the future but rather about restoring an idealized and romanticized golden age that was lost in the past (Hutcheon 1998). This reactionary nostalgia features prominently in American utopias since they are but a product of the Puritan rhetoric that still shapes the self-perception of the dominant cultural groups in the U.S. There is a pronounced utopian thread woven together with the narrative fabric of American exceptionalism. In short, it is difficult to make the American city in general, and the city of the visual SF imagination in particular, hold still for any length of time. Hence, any taxonomic appraisal of urban geographies must by necessity be rooted in a firm conceptual understanding of the American city.

Conceptualizing and re-historicizing the American city of the (SF) imagination

The pioneering work of film scholar Vivian Sobchack lends itself to a working definition for the imagined American city in visual SF. She defines it as a 'hypnogogic site', which is 'more than mere background' since it provides 1) 'a significant and visibly signifying shape', and 2) 'a temporal dimension, a historical trajectory', which is oriented towards the future (Sobchack 78). In her work, she charts this trajectory from the promise and failure of modernism's aspirations in the form of images that document the destruction and desertion of urban environments to images of 'urban exhaustion, postmodern exhilaration and millennial vertigo' (Sobchack 78). Unfortunately, the scope of Sobchack's city mapping is strictly limited to 'the imaginary city of the American science fiction film from the 1950s to the 1990s' (Sobchack 78). Her elaborations only start by looking at images that speak to the failure of the modern(ist) city. Moreover, while acknowledging that 'detached images of the American science fiction film city ... are not be be seen as ahistorical or absolute and essential', the phenomenological history she aims to offer quickly turns into a chronology that gives preference to visual aesthetics with only occasional nods to cultural context(s), eschewing the historio-cultural roots of the American city of the (SF) imagination altogether (Sobchack 79). The 'poetic image' of the American SF city remains front and center (Sobchack 79).

Even so, Sobchack's definition, observations and conclusions also point to a *quintessential dualism* that lies at the heart of the American city's palimpsestuous character; it had shaped American thoughts about and imaginings of the city long before the progressive aspirations of modernism were architecturally re-inscribed in the city of the future at the turn of the 19th century. Re-historicizing the SF city will allow for a deeper appreciation of this highly visual(ized) and readily imagined *topos* by probing beyond its surfaces which tend to dazzle and distract by triggering a mix of awe and anxiety. In order to do that it serves to harness one of the generic and intellectually stimulating devices of SF. Less constrained by their conventions, SF along with other fantastic genres are enabled to engage differently

162

with that which is normally confined to the abstract, metaphorical or subjective. The narrative freedom and thought-provoking power of SF coalesce in one of its rhetorical devices: the literalization of metaphor. In short, while an utterance such as "her world exploded" can only ever bear metaphorical meanings in realist fiction, it might well have to be taken and understood as a literal occurrence/event—thus being hyper-realized—within the fantastic framework created in SF. In the American context, the dualism that Sobchack points to has repeatedly been expressed in diverse cultural artifacts prior to the modern(ist) moment. Consequently, the SF city can be seen to literalize the metaphors and values attributed to the images and imaginings of the American city as a New Jerusalem and its Jeffersonian counter-image.

Upon arriving in the "New World", John Winthrop, Puritan leader and first governor of Massachusetts, famously imagined the colony's works as 'a city upon a hill' in a sermon he delivered to his followers (Baym and Levine 177). Winthrop's city image is constructed as an exceptionalist narrative which sees the Puritans—god's chosen people—carve out a New Jerusalem, redeeming it from the hands of the "savage". After all, following Puritan rhetoric, god had set the New World aside for the arrival of his chosen nation. He preached that as such '[t]he eyes of all people are upon us … we shall be made a story

163

and a by-word through the world' (Baym and Levine 177). The image of the city was inscribed not only in their colony, but also became the locus of meaning through which god's promise was to be redeemed for America as a whole by way of their descendants' accomplishments. Winthrop's assertion that their city's story would be closely followed by the entire world, found articulation in the two themes—one exemplary, one missionary—that together fuel American exceptionalism.

The image of the city as a place where America's guaranteed success story would play out, also persisted. Three hundred years after Winthrop, William James saw in the modern American city 'the center of the cyclone', which he thought was 'magnificent' since 'the great pulses and bounds of progress' point 'in directions all simultaneous that the coordination is indefinitely future' (James 67). Even so, being under the scrutiny of the entire world, Winthrop also summoned a latent specter of the biblical Sodom and Gomorrah lest the Puritan mission falls short of redeeming god's promise; dealing 'falsely with our God in this work we have undertaken', would 'cause Him to withdraw his present help for us' (Baym and Levine 177). Consequently, what the American city—and America-as-a-city—should become, always inevitably includes that which it should not become. It has the seeds of decay and failure already built-in which germinated most clearly in the Jeffersonian pastoral.

For two hundred years prior to the founding of the U.S., the images and aspirations projected onto and coming out of the New World had actually been rural and anti-urban. A pastoral vision for America's future then also informed the rapid westward expansion that followed in the wake of independence. Ironically, it obscured the decidedly (proto)industrial and urban-based power structures that actually drove the westward march of the American empire. Thomas Jefferson arguably did the most to intellectualize the agrarian vision for the U.S. which aimed to lead away from entanglements with the "Old World" and towards self-sufficiency, economic independence and ultimately supremacy. In the process, he bolstered his pastoral vision with a sharp critique of urban environments, adding a call for his fellow citizens to abjure the corruptive effects of the city.

In a letter to Benjamin Rush, Jefferson stated that cities are 'pestilential to the morals, the health and liberties of man' (Meagher 81). He had already articulated his revulsion for cities earlier in his *Notes on the State of Virginia* (1787). In his chapter on 'manufactures', he offers that '[t]he mobs of great cities add just so much to the support of pure government, as sores do to the strength of the human body' (Jefferson 217). For him, '[t]hose who labour in the earth are the chosen people of God' not least because the '[c]orruption of morals in the mass of cultivators is a phenomenon of which no age nor nation has furnished an example' (Jefferson 217). Ironically, his view is more aligned with the historical realities confronted by the Puritans since their 'city upon a hill' was actually a colony that subsisted on communal farming.

While having undoubtedly partaken the offerings of a city like Paris while serving as ambassador to France, he advocated for an anti-urban ideology for the newly formed republic. For Jefferson, the threats of urbanism were twofold: 1) the city jeopardized the virtues of the individual, and by the same token 2) it endangered the democratic fabric of the republic itself. Having been articulated much closer to the emergence of the modern(ist) American city, Jefferson's apprehensions also persisted and continued to inform its palimpsestuous character.

For example, Reverend Josiah Strong reiterated them in *Our Country* (1885); his city was a 'storm center', where 'wealth is massed ... the sway of Mammon is widest', and where 'the ennui of surfeit and the desperation of starvation ... raise riots for the purpose of destruction and plunder; here gather foreigners and wage-workers ... especially susceptible to socialist arguments' (Campbell and Kean 190-191).

The discourses that constitute both city images equally informed the emergence of the modern American city of the future at the end of the 19th century. Hence, the palimpsestuous character of the American city, which became increasingly stratified, is contradictory at worst, and ambivalent at best. In order to conceptualize it along with the imagined American city in visual SF, it serves to turn to social theorist Michel De Certeau. The exponential growth of urban areas and the deteriorating living conditions only seemed to confirm Jefferson's doubts. This translated into attempts to reinvent and properly *plan* the city by way of imposing a socio-scientific order upon it. Architects, city planners, and social progressives led the charge which arguably did more to formalize the historical dualism than to dissolve it. De Certeau distinguishes between the city as a 'picture', or concept, and the city as 'practices' (Campbell and Kean 194).

The former speaks to viewing and reading the city from a high vantage point such as a skyscraper, an airplane, or from the drawing board of city planners. Imposed from above by the tower of corporate power—a Foucauldian panopticon—we see the totalizing and utopian discourses of ordering, regulating and reading the city (Collie 426). Yet, it

167

reveals only the 'theoretical' city as it is coerced into the conformity of the grid (Collie 426). For the city-as-practice, we need to descend onto the street level which eludes the controlling gaze of the panopticon. There we find chaotic and fractal interactions and exchanges that defy the conformity of the grid at the sites of popular culture which cannot be easily observed from the remove of a high vantage point (e.g. restaurants, movie theaters, shops, workplaces). These 'intersecting writings compose a manifold story that has neither author nor spectator' (Collie 426). De Certeau's vantage points not only allow for visualizing the aesthetics of the modern(ist) American city, but also for understanding the underlying ideological and cultural forces that shape its "surfaces".

The moment of the modern(ist) American city and the genesis of the city in visual SF

During the *fin de siècle*, American cities became the epicenters of modernity which coincided with a peak in utopian aspirations that was fueled by progressivist ideas. The modern(ist) American city became the place where the possibilities of new technologies, industrialization and commerce coalesced in a rapidly densifying and localized space. It brought people and their ideas together in a mix of productive art and literature that reflected on the

168

urban impulse. At the same time, a religious revivalist movement swept through the country which saw many religious groups form their own communes that aimed to achieve utopia, i.e. heaven on earth. According to Lincoln Geraghty, these communal experiments where 'characteristic of the nation's drive for religious and cultural reform' (Geraghty 4). While diverse in practice, these communities shared the belief that utopia could and would be realized on earth by god's chosen people; and, utopia was destined to be "made in the U.S. of A".

Even though the utopian impulse was largely rural at first, it soon blended with city-based progressive movements that sought to righten the social and economic wrongs of the Gilded Age; wrongs that were brought to light by muckraking authors such as Upton Sinclair or Lincoln Steffens in his novel *The Shame of the Cities* (1904).

Nowhere was the vision of a future urban utopia presented more clearly than in the quintessential utopian novel of the time, Edward Bellamy's *Looking Backward* (1888). Its time traveling protagonist, John West, awakens in Boston in the year 2000 and finds himself exploring the following cityscape: '[m]iles of broad streets, shade by trees an lined with fine buildings ... stretched out in every direction. ... Public buildings of a colossal size and architectural grandeur unparalleled in my day raised their stately piles on every side' (Geraghty 5). Bellamy celebrates the teleology of progress which guarantees America's march into the future—an urban future.

The Boston of his future showcases how scientific socialism is put to work in order to achieve equality of labor by *steering* people away from excessive individualism and towards communal thinking and the centrality of public life through urban planning. However, Bellamy's urban utopia is decidedly undemocratic, bordering on totalitarianism, and exceedingly dependent on automatization and mechanization. Replicating the palimpsestuous dualism outlined earlier, Andrew Wood asserts that Boston like 'most idealized forms of public life hide a machine under their well-tended gardens' (Wood n.p.). Even so, the novel furnished architects, urban planners and progressive policymakers with an impetus and a visual as well as an ideological grammar for designing the American city of the future.

Bellamy's descriptions of a utopian American urban future ultimately informed and translated into the architecture of the World's Columbian Exposition in Chicago in 1893. Fueled by and feeding further into the City Beautiful Movement, the goal of the world's fair was to present a model for urban beautification and monumental grandeur that was achieved through American exceptionalism and its reciprocal relationship with technological progress. Combining Beaux-Arts architecture with the power of the latest technologies, especially electricity, the expo showcased the "White City" as a model for the future. It featured modern public transport systems and the commercial amenities of a new century while signs of poverty and pollution were visibly absent. Applying the discursive tools of corporate power and organization, Alan Trachtenberg argues, planners and architects such as Daniel Burnham and Frederick Law Olmstead aimed at a 'benign coercion ... to eradicate the communal culture of working-class and immigrant streets, to erase that culture's offensive and disturbing foreignness, and replace it with middleclass norms of hearth and tea table' (Campbell and Kean 194).

In the vein of De Certeau, these "benevolent" futurists dreamt of using the city-as-concept to contain, control and thus legitimize the city-as-practice. This actually gave rise to and formalized a clash between 'civil horizontalism and corporate verticality' (Campbell and Kean 194). The tower, and later the skyscraper, became their residence and the symbol of the Foucauldian power which they directed at the streets below from the remove of the panopticon.

When these modern(ist) cityscapes are visually imagined, what comes to mind is rather familiar since Bellamy's novel and the Chicago expo foreshadowed the imagined cities of the future in SF films of the 1920s and 30s. First and foremost, there is *Metropolis* (Lang 1927). Lang was inspired by a visit to New York.

He also drew on Bauhaus architecture and borrowed heavily from American futurists such as Hugh Ferris and Raymond Hood who, in turn, had been influenced by H. G. Wells' vision of an inflated New York City in *The Future in America* (1906). Being key to the 'modernist vision' encoded in the emerging city of the SF visual imagination, Mike Davis argues that Wells, just like Lang, already tried to 'envision the late twentieth century by enlarging the presented ... to create a sort of gigantesque caricature of the existing world, everything swollen up to vast proportions and massive beyond measure' (Davis n.p.). However, *Metropolis* not only gave us 'the look of the future'—its modernist iconography—but also, like in Bellamy's novel, a 'reordering of society' that is 'based on city planning, social theory, national and economic reform' (Geraghty 5).

As an imaginary laboratory, it spoke to the reorganization of society by way of rationalizing the labor system in the vein of Henry Ford. With the

172

above ground city emphasizing 'transport, entertainment, and consumption', its seeds of decay are revealed on 'its subterranean levels' which are 'home to an enslaved underclass existing only to keep the city running for the elite class living above' (Geraghty 6).

With *Metropolis*, the palimpsestuous character of the American city as the New Jerusalem and the New Sodom and Gomorrah was also carried over and effectively installed in visual SF. It was consolidated visually and narratologically in films like *Just Imagine* (Butler 1930) and *Things to Come* (Menzies 1936). In these movies, the city of the SF imagination continues as a hierarchy of corporate and capitalist power where a large labor force ensures that that goods are shipped to the higher echelons of the urban utopia ready for consumption. *Just Imagine* is an interesting case not least because it weds SF futurism with musical comedy.

Moreover, it reverberates only with the vertical power, vast size and etherial delicacy of Lang's upper city, eclipsing its dark abyss completely. Sobchack opines that 'however quotidian, [the city] is nonetheless emphatically and literally uplifted' (Sobchack 79). *Just Imagine* already directs us to this latent modern(ist) semblance of utopia that would continue to inform the many different urban spaces that have been imagined in SF since then. The reciprocal relationship between utopian literature, urban planning ventures and progressivist ideas remained alive and well for years to come.

These early utopian cityscapes of the SF imagination fed into the urban visions that were presented at two more world fairs in New York in 1939 and 1964, respectively. They clearly prefigured the designers' updated modernist vision as both expos celebrated *the world of tomorrow* for transportation, entertainment and shopping. Their urban visions were extrapolated from the expansion of the automobile age, the introduction of the television, and later the space age. For example, it was no coincidence that the world's fair of 1939 served as a frame for the first world science fiction convention, Nycon 1.

The architectural icons of the expo—the Trylon, the Perisphere and the Helicline—still shape the cityscapes in visual SF today in one permutation or another; Disney's *Tomorrowland* (Bird 2015) is one of the most recent examples. At the fair in 1964, many structures were built in a mid-century modern style that was heavily influenced by Googie architecture. It was in turn influenced by the growing car culture, design elements of jet aircraft, and a large dose of the space and atomic age. The TV-series, *The Jetsons* (1962-63) is a contemporary example that used this style in an SF format. More recently, the post-apocalyptic mis-en-scène in the *Fallout* video game series is populated by the architectural remnants of the atomic age. At both fairs, large corporations, such as General Motors and General Electric, put futuramas on exhibit in which visitors could step into the urban world of the near future; for many of them it must have felt like stepping onto the set of the latest SF movie. For example, a few years later, the urban environment in the film *Logan's Run* (Anderson 1976) is reminiscent of the futuristic visions presented at the two fairs. The reciprocal relationship between the utopian aspirations of American modernism and world fairs infused the American city with sufficient substance to serve as an SF *topos* that could react to new fears such as a nuclear fallout, overpopulation, ecological disasters and terrorism. It also became a common place where new technological developments, especially the digital revolution, would install readily recognizable cities of the imagination in visual SF.

A taxonomic tale of four cities
In the past, most critical attention to cities of the imagination was geared towards SF films which ultimately also informed the *stringently* dualistic taxonomy that was used to categorize these spaces. The "city of light" versus the "dark city" became the dominant paradigm as evidenced in Vivian

175

Sobchack's work. Jonathan Raban reiterated that the 'the soft city of illusion, myth and aspiration', is in this constant tension with the 'hard city' of nightmare (Collie 427). As such, it has arguably obscured the potential for a more precise taxonomic appraisal of urban geographies in visual SF not least because this dualism speaks to the highly visual(izing) aesthetics of the city *topos*. While there was a long stretch of particularly dystopian urban scenarios that razed the flawed utopian promise of the American city in every imaginable way between the 1950s and late 1980s, all of the resulting types of imagined cities still populate SF texts today because they all contain a latent semblance of the original modern(ist) urban utopia—the city-as-concept. Since the early 1990s, visual SF has seen a constant replay and rehash of already established urban geographies which are simply adapted to reflect new social aspirations and new fears. As showcased by the genesis of the city of the future in visual SF, the continued interrelationship between imagined SF cities and contextual impulses points to a segway that allows for a more diversified, four-tiered taxonomy.

Real world impulses for SF cities and the resulting impromptu taxonomic overview can be structured as follows: the modern(ist) promise of the city-as-concept continues to resonate in 1) *urban utopias*, carrying both the aesthetics of world fairs and utopian literature, and their underlying ideology, well into the 21st century. In the wake of Hiroshima, 2) *urban wastelands*, which may or may not be postindustrial or post-apocalyptic began to speak to fears of dissolution and invasion. Such SF cities often feature the destruction of known landmarks, images of emptiness or, alternatively, crowdedness. In short, they are beset by the growing threat of disasters regardless of whether they are man-made and/or natural. The prolonged echo of the space race, especially the groundbreaking work on space habitats by American physicist and space booster, Gerard K. O'Neill, nurtured and sustained imagined 3) *sky cities and cities in outer space*, which are often overlooked in critical approaches to urban spaces in SF.

Lastly, it is not difficult to see how the exponential growth of digital data networks, ranging from GPS/satellite mapping, public transportation maps and the internet to CCTV surveillance, geo-fencing and virtual reality, has fed into a proliferation of 4) *virtual(ized) cityscapes* in SF since the early 1980s. Once each city type was imagined, they become readily accessible at any time. Hence, while they emerged more or less chronologically in response to changing contextual impulses, they now all co-exist simultaneously in the SF imaginarium. These urban imaginings often overlap and cross-pollinate.

Urban utopias are hubs or centers of corporate, capitalist and/or imperialist power that always bespeak either a literal or metaphorical underside, i.e. a darker, disenfranchised space that exists side by side with its shining counterpart. This type of SF city essentially speaks to the ongoing sequence of permutations of the *Metropolis*-archetype which has never really gone out of fashion in visual SF.

For example, *Logan's Run* offers us such a garden-variety of a domed—and doomed—urban paradise. Controlled by a computer, the city has one major drawback. The inhabitants may enjoy its hedonistic life only until they turn thirty. That is when they undergo the ritual of the 'Carrousel' and are disintegrated. Individual freedom and the chance to grow old are found outside this flawed paradise in the ruins of what used to be Washington, DC. *Star Trek: The Original Series* (1966-69) offers similar scenarios. In the first-season episode 'A Taste of Armageddon' (Pevney 23 Feb 1967), the *Enterprise* visits two urbanized civilizations who have been engaged in a virtual "hot" Cold War for 500 years. In order to preserve the semblance of their utopian civilization and their achievements, attacks are calculated and carried out electronically. Citizens are voluntarily vaporized in death booths whenever the enemy scores a hit. The episode 'The Cloud Minders' (Taylor 28 Feb 1969) is perhaps even more obvious in that it is only a vaguely disguised adaptation of *Metropolis*. Captain Kirk and his crew travel to Merak II on a mission to obtain a rare mineral from an alien people who reside in a lavish cloud city called Stratos. They soon become embroiled in a terrorist attack by the Troglytes, who are revealed as the enslaved labor force that mines the precious mineral in a toxic environment on the planet's surface.

More recently, the movie *In Time* (Niccol 2011) offers us an urban geography where the slick, vertical, corporate center, which controls the flow of temporal capital, is strictly segregated from the horizontal proletariat outliers. This urban geography enables a narrative that allegorizes the Occupy Wall Street movement. For a more nuanced representation of the same urbanized conflict, one can turn to the time travel/crime procedural *Continuum* (2012-15). It is partially set in a near future where the North American Union is governed by an oligarchic and transnational corporatocracy, the Corporate Congress. One of the main players, SadTech, is based in Vancouver from where its corporate tendrils feed off exploitative labor farms that are located beyond the outskirts of the city. The seemingly utopian city life is strictly regulated by a corporate police force, the City Protective Services (CPS). Lastly, the utopian albeit crowded metropolis of the United Federation of Britain, having swallowed American cities wholesale in the remake of *Total*

Recall (Wiseman 2012), subsists on labor power which is imported quite literally from the other side of the globe—an urban wasteland called 'The Colony'.

Urban wastelands abound in visual SF and they vastly outnumber other types of SF cities. Since the 1950s, the flawed utopian promise of the American city has been visually exposed and destroyed countless times. Initially, urban geographies in SF films served to react to fears of a possible nuclear fallout and/or Soviet invasion. Films such as *When Worlds Collide* (Maté 1951), *Earth vs the Flying Saucers* (Sears 1956), or *On the Beach* (Kramer 1959), always feature clearly identified American cities and their landmarks that are subsequently exposed to a 'destructive force' which may take one of three forms, i.e. 'an apocalyptic natural force', 'a primal Best or Creature', or 'a technologically superior alien war machine' (Sobchack 81). *Urbicide* was the dominant theme. These scenarios became so ingrained in the visual SF imagination that they have recently been resuscitated with great vigor in the alien invasion TV-series *Falling Skies* (2011-15), and as an embedded narrative in the *Fallout* video game series. The latter comes complete with a post-nuclear re-imagining of 1950s popular/everyday culture as the player character explores the urban wastes of Washington, DC, Las Vegas and Boston. These urban wastelands aim to trigger a strong affect with the viewer/player by emasculating and/or completely leveling known landmarks, i.e. the pillars of American civilization. For an earlier example, the half-buried Statue of Liberty at the end of *Planet of the Apes* (Schaffner 1968) comes to mind.

Images of stillness and emptiness also prevail, and they set protagonists/player-characters in stark contrast to their environments. The movie *I am Legend* (Lawrence 2007) and the recent TV-series *The Last Man on Earth* (2015–) serve as potent examples. The announcement trailer for the latter portrays this vividly; it shows Phil Miller singing the *Star-spangled Banner* in an empty Dodgers stadium. The camera gradually recedes to reveal Los Angeles being completely devoid of human activity, interspersing it with cuts to other urban centers like London or Kuala Lumpur only to show similar scenes of emptiness. These images clearly aim to trigger a sense of awe.

In the wake of fears about totalitarian tendencies in the American government, overpopulation, food shortages, and pollution starting in the late 1960s, SF cities of the future also gave us the opposite of emptiness. For example, *Soylent Green* (Fleischer 1973) envisions a future New York City that 'no longer aspires but suffocates and expires' (Sobchack 82). Vertical power and control have lost any semblance of meaning and relevance. The same goes for Mega City One in *Judge Dredd* (Cannon 1995), including its recent remake *Dredd* (Travis 2012). In these urban wastelands, the view from above usually produces only a weak image of the city-as-concept, if any at all. These cities are simply too dense. A sense of being lost and disoriented already permeates higher vantage points. The gaze from the Foucauldian panopticon is greatly weakened. This is why images of the city-as-practice predominate. Next to *Metropolis*, *Blade Runner*'s (Scott 1982) Los Angeles of 2019 is arguably the second most iconic and paradigmatic cityscape in visual SF. Even though it is often celebrated as a postmodern pastiche *par excellence* that points us to the inevitable future of urban decay, Mike Davis invites us to take a closer look. Once we remove Ridley Scott's and Syd Mead's dark, 'Yellow Peril', and noirish veneer of the city, what remains, he argues, is 'recognizably the same vista of urban gigantism that Fritz Lang celebrated in *Metropolis*' (Davis n.p.). Once again it underscores how the palimpsestuous character of the American city is literalized by SF urban geographies.

Sky cities and *cities in outer space* have often been overlooked in most histories of SF cities (Westfahl 92-103). The floating city of Columbia, which lent its name to this essay's title, is the main setting of the recent successful video game *Bioshock Infinite* (2013); and it serves as a levitating waypoint. The city is a steampunk/retro-futurist critique of the historical moment of the modern(ist) American city which was outlined earlier. It is no coincidence that Columbia was completed in 1893 and launched at the Chicago World's Fair. As a *topos*, it comes embedded with a deeply troubled vision of secessionist utopianism and religious fanaticism. In SF literature, cities in outer space usually serve as pastoral, or at least suburban opposites for overcrowded and polluted urban planet surfaces. In visual SF, this is not so much the case in part because the most iconic of these cities—which are not that numerous—are actually large-scale military/government installations.

They can also be visualized along the lines of De Certeau. For example, there is the view from one of the docking pylons down onto the space station of *Deep Space Nine* vis-à-vis its Promenade. It can be compared to the view offered to a visitor traveling on *Babylon 5*'s Maglev vis-à-vis its main market street, the Zócalo. Both city stations offer much darker spaces that are located on lower levels, such as the Lower Core, or Brown and Grey Sectors, respectively. In the *Mass Effect* (2007-12) video game trilogy, the Citadel is the galaxy's central governmental metropolis, featuring little pastoralism and offering ample opportunities to get into trouble on the darker levels of the Wards. It serves to remember that most of these cities in outer space are analogous to colonial sites. They fulfill the function of port cities in what are wordbuilding paradigms which are modeled on pre-industrial/early modern historical periods. *Deep Space Nine* was actually intended to be like an exotic port in the West Indies during the late 18th century.

Virtual and *virtualized cityscapes* have often been segregated as an aberration from other urban environments in visual SF. While they indeed challenge the viewer/player, it is essential to include them in the same critical framework not least because they maintain a mimetic link to physical urban spaces. Moreover, despite their fluid and rather abstract characteristics, the virtual(ized) city and especially 'Cyberpunk's vision of near-future cities often gels well with formal urban theory' (Collie 428). Collie stresses that it 'highlights the central role of global cities in world economic and social change; the development of cities as communication systems; the dominance of Pacific Rim economies and culture; intense bifurcation and absent or in-crises middle class; the weakening role of government; and the importance of Los Angeles as a model for our urban future' (Collie 428).

Even so, like it is the case with space stations, there is a noticeable difference between virtual(ized) cityscapes in SF literature and visual SF. Ever since the digital revolution, the city has extended its physical limits virtually *ad infinitum* in that, according to Davis, it 'redoubled itself through the complex architecture of its information and media networks' (Collie 428). One only has to think of the maps of transportation systems, traffic and CCTV surveillance, GPS mapping, the multiplying layers of social media spaces, and Alternate Reality Games (ARGs) in order to understand that cities are no longer contained by as well as in three-dimensional space. Since the rise of Cyberpunk literature, and *Neuromancer* (1984) in particular, we have access to Gibsonian maps for the cyberspace city as 'a simulation of the city's information order', where vertical power structures have been converted into virtual conduits of capital and information (Collie 428).

These cities are postmodern topographies where hackers 'stroll through the luminous geography of this mnemonic city where data-bases have become 'blue pyramids' and 'cold spiral arms'', and the sky has the color of television noise (Davis n.p.). Even though Cyberpunk cities transcend most if not all boarders of the physical world, and are thus able to conjure up images that are radically different in the reader's mind, the way how virtual cityscapes are visualized is still circumscribed by an iconography of familiarity. In other words, while the Chicago world's fair and its modern(ist) vision of an urban American future were undoubtedly ahead of their time, they were unable to anticipate the city as simulation and its subsequent virtualization. This lack of conceptual space still lingers in visual SF.

Since virtual cities are governed by the flow of capital and information, we see in a sense a digital corporate wasteland when we look at films like *Tron* (Lisberger 1982) and *Tron: Legacy* (Kosinski 2010). Their urban geographies are still ordered along grid-like dimensions that are found in the non-virtual world. Governed by a Foucauldian panopticon—a virtual echo of the tower—these urban spaces ultimately give us a hyper-real(ized) version of De Certeau's city-as-concept. Similar observations can be made about virtual cites that are less abstract(ed), such as Mega City in the *Matrix*-trilogy (Wachowski and Wachowski 1999, 2003), or the diesel-punkish/noirish geography of New Cap City in the TV-series *Caprica* (2010). They confirm that when it comes to the virtualized American city in SF, it seems that we are unable to visualize these spaces without radically breaking with mimetic constraints.

Conclusion: American cities at rest?

The taxonomic re-appraisal, re-mapping and re-historicizing offered in this essay might have seemed like a ride on a futuristic subway as utopian, dystopian, outer-space and virtual urban geographies of the visual SF imagination swooshed by. Even so, it has made clear just how resilient the flawed utopian promise of the modern(ist) American city really is in that it continues to shape the literalization of its palimpsestuous character in visual SF. It serves to remember that '[t]raditionally America's spatial mythology has privileged the non-urban, and has been, indeed, anti-urban' (Sobchack 81). The paradise supposedly found in the New World was symbolically located on the frontier, the wilderness of the west, where it was to be redeemed by god's chosen nation. This process was deeply ironic since expansionist power structures were firmly rooted in the industrial urban centers of the east. At the historical moment of the modern(ist) American city, Jonathan Winthrop's powerful metaphor of America as 'a city upon a hill' was first literalized and then continuously problematized in the cities of the SF imagination which emerged from the intersection of utopian thought, progressive ideology and city planning.

At the same time, Thomas Jefferson's vision for a pastoral nation, and the harsh criticism of urban spaces it included, still haunts the American city in general and the cities in visual SF—especially the urban wastelands. Even so, the four types of the SF city inevitably point to the resilience of the utopian promise inherent in modern(ist) American city, and the people who built and live in it. Marshall Berman emphasizes that the American people 'have the

187

capacity both to commit urbicide and to overcome it; to reduce their whole civilization to ruins, but also to rebuild the ruins' (Campell and Kean 212).

The city in the eponymous TV-series *Defiance* (US 2013-15) makes for a compelling and concluding case in point. In the show, earth becomes the new home for alien refugees, the Votan Collective, who consist of eight different species. Their arrival first triggered a global conflict which was followed by the destruction of their ark fleet. Countless pieces of debris crashed on the surface leading to the radical terraforming of earth's geo-/biosphere.

The series is set in a growing township which is located in a place that used to be St. Louis. The surrounding area is not only populated by a hostile flora and fauna, but it also offers rich mineral deposits and energy sources. The town of Defiance, where humans and Votans co-exist, continues to thrive underneath the Gateway Arch, which, though battered, miraculously survived terraforming while 'Old Saint Louis' became the subterranean basis for the renewed growth of the settlement.

Though threatened by an alien wilderness and caught between two political fronts—the mildly fascist Earth Republic and the somewhat reclusive Votanis Collective—the Arch, designed by neofuturist architect Eero Saarinen in the late 1940s early 1950s,

188

once again speaks to the archetypal resilience of the promise inherent in the American city. Just like the Arch, the American city, America-as-city, and the city of the visual SF imagination remain *defiant*.

Works cited
Print

Abbott, Carl. 'Cyberpunk Cities: Science Fiction Meets Urban Theory'. *Journal of Planning Education and Research* 27 (2007): 122-131.

Campell, Neil, and Alasdair Kean. *American Cultural Studies: An Introduction to American culture*. 3rd ed. New York: Routledge, 2012.

Collie, Natalie. 'Cities of the imagination: Science fiction, urban space, and community engagement in urban planning'. *Futures* 43 (2011): 424-431.

Davis, Mike. 'Beyond Blade Runner: Urban Control - The Ecology of Fear'. *Mediamatic Magazine* 8.2/3 (1995): n.p.

Geraghty, Lincoln. *American Science Fiction Film and Television*. Oxford: Berg, 2009.

James, William. 1920. *The Letters of William James, vol II*. Ed. Henry James. Boston: Houghton Mifflin, n.d.

Jefferson, Thomas. 1787. *The Portable Thomas Jefferson*. Ed. Merrill Peterson. New York: Penguin Books, 1977.

—. 1787. 'Manufactures'. *Philosophy and the City*. Ed. Sharon M. Meagher. Albany: State University of New York Press, 2008. 81-83.

Sobchack, Vivian. 'Cities on the Edge of Time: The Urban Science Fiction Film'. Ed. Sean Redmond. *Liquid Metal: The Science Fiction Film Reader*. New York: Wallflower Press, 2007. 78-87.

Westfahl, Gary. *Islands in the Sky: The Space Station Theme in Science Fiction Literature*. 2nd ed. Rockville: The Borgo Press, 2009.

Winthrop, John. 1630. 'A Model of Christian Charity'.
    Ed. Nina Baym and Robert S. Levine. *The Norton
    Anthology of American Literature: Volume A:
    Beginnings to 1820*. 8th ed. New York: W. W.
    Norton, 2012. 166-177.

Web

Hutcheon, Linda. 'Irony, Nostalgia, and the
    Postmodern'. *Library.utoronto.ca*. 1998.
    <http://www.library.utoronto.ca/utel/criticism/hut
    chinp.html>. Accessed 7 Jan 2016.
Wood, Andrew. 'Looking back to the Year 2000'.
    Sjsu.edu. 2000.
    <http://www.sjsu.edu/faculty/wooda/s149/149syll
    abus6.html>. Accessed 8 Jan 2016.

Film

Blade Runner. Dir. Ridley Scott. Warner Bros., 1982.
Dredd. Dir. Pete Travis. Lionsgate, 2012.
Earth vs the Flying Saucers. Dir. Fred Sears. Columbia
Pictures, 1956.
I am Legend. Dir. Francis Lawrence. Warner Bros., 2007.
In Time. Dir. Andrew Niccol. Twentieth Century Fox,
2011.
Judge Dredd. Dir. Danny Cannon. Buena Vista Pictures,
1995.
Just Imagine. Dir. David Butler. Fox Film Corporation,
1930.
Logan's Run. Dir. Michael Anderson. United Artists,
1976.
Metropolis. Dir. Fritz Lang. UFA, 1927.
On the Beach. Dir. Stanley Kramer. United Artists, 1959.

Planet of the Apes. Dir. Franklin Schaffner. 20th Century Fox, 1968.

Soylent Green. Dir. Richard Fleischer. Metro-Goldwyn-Mayer, 1973.

The Matrix. Dir. Andrew and Lana Wachowski. Warner Bros., 1999.

The Matrix Reloaded. Dir. Andrew and Lana Wachowski. Warner Bros., 2003.

The Matrix Revolutions. Dir. Andrew and Lana Wachowski. Warner Bros., 2003.

Things to Come. Dir. William C. Menzies. United Artists, 1936.

Tomorrowland. Dir. Brad Bird. Disney, 2015.

Total Recall. Dir. Len Wiseman. Columbia Pictures, 2012.

Tron. Dir. Steven Lisberger. Buena Vista, 1982.

Tron: Legacy. Dir. Joseph Kosinski. Disney, 2010.

When Worlds Collide. Dir. Rudolph Maté. Paramount Pictures, 1951.

Television
Caprica. Syfy. 2010.

Continuum. Showcase. 2012-15.

Defiance. Syfy. 2013-15.

Falling Skies. TNT. 2011-15.

House of Cards. Netflix. 2013–.

"A Taste of Armageddon." *Star Trek: The Original Series*. NBC. Dir. Joseph Pevney. 23 Feb. 1967.

"The Cloud Minders." *Star Trek: The Original Series*. NBC. Dir. Jud Taylor. 28 Feb. 1969.

The Jetsons. ABC. 1962-63.

The Last Man on Earth. Fox. 2015–.

Video games

Bioshock Infinite. Irrational Games. 2013.

Fallout 2. Black Isle Studios. 1998.

Fallout 3. Bethesda Softworks. 2008.

Fallout: New Vegas. Bethesda Softworks. 2010.

Fallout 4. Bethesda Softworks. 2015.

Mass Effect. BioWare. 2007.

Mass Effect 2. BioWare. 2010.

Mass Effect 3. BioWare. 2012.

The Last of Us. Naughty Dog. 2013.

Chapter 8. Mythology in science fiction. [3]– Mariella Scerri and David Zammit

Introduction

Since the beginning of humankind's existence, myths have functioned as rationalizations for the fundamental mysteries of life. In the absence of scientific information of any kind, societies devised creation myths, resurrection myths and complex systems of supernatural beings, each with specific powers, and stories about their actions. Since societies were often isolated from each other, most myths evolved independently. Nonetheless, the various myths are surprisingly similar (Dundes 2).

The need for myth is a universal need. Over time, one version of a myth would become the accepted standard that was passed down to succeeding generations first through story-telling, and then, much later set down in written form. Inevitably myths became part of systems of religion, and were integrated into rituals and ceremonies, which included music, dancing and magic.

---

[3] This piece was originally published in the SFRA Review as Scerri, Mariella and Zammit, David. "Mythology in Science Fiction ." *Science Fiction Review Association.* 316 (2016): 15-21.

In modern society, myth is often regarded as historical or obsolete. Many scholars in the field of cultural studies are now beginning to research the idea that myth has worked itself into modern discourses (Bascom 20). Modern formats of communication allow for widespread communication across the globe, thus enabling mythological discourse and exchange among greater audiences than ever before. Various elements of myth can now be found in cinematography and video games. Although myth was traditionally transmitted through the oral tradition on a smaller scale, the technology of the film industry has enabled film makers to transmit myths to large audiences via film dissemination (Singer 5). Film is ultimately an expression of the society in which it was credited, and reflects the norms and ideas of the time and location in which it was created. In this sense, film may be regarded as the evolution of myth. Koven opines that the basis of modern story telling in both cinema and television lies deeply rooted in the mythological tradition.

Many contemporary and technologically advanced movies often rely on ancient myths to construct narratives (Koven 180).

Mythological archetypes such as the cautionary tale regarding the abuse of technology, and battles between gods during creation stories are often the subject of major film productions. Hodge further makes claims that 'the heroic figures of today's fantasy and science fiction are merely the latest in a long line of culture heroes who purport to be models of all that is best in our society and thus offer a comforting example of how truth and goodness prevail against evil and lies' (Hodge 37). In an interview on science fiction and mythology, Grech contends that science fiction gave a paradigm shift to the world of ancient mythology (Fava). Contemporary television programmes and books with science fiction and fantasy story lines employ recreations of characters, incidents and motifs from the great 'oral literary' heritage (Hodge 38).

The following paper will broadly explore the presentation of mythology in science fiction books and television series. The *Keltiad* by Kennealy Morrison and *The Ilium* and *Olympos* by Dan Simmons will be considered while selected relevant episodes from *Star Trek* and *Stargate* will be discussed such that myth representation will be analyzed.

Mythology Representation in Science Fiction
The relationship of mythology to science fiction is not always obvious. Blish asserts that this confusion stems from the widely held belief that science fiction

is itself a form of latter day mythology, fulfilling comparable hungers in us. Blish points out that myth is usually 'static and final in intent and thus entirely contrary to the spirit of science fiction, which assumes continuous change (Blish).

Traditionally mythology appears in science fiction in two ways, its archetypes being either re-enacted or rationalized. The re-enactment of myths is the more complex of the two cases. Behind the retelling of a myth in a modern context lies the feeling that, although particular myths grew out of a specific cultural background, the truths they express relate to the humanness and remain relevant to all societies. The story of Prometheus, punished by the gods for stealing fire from the heavens, or its Christian variant, where Dr Faustus is doomed to external damnation for selling his soul in exchange for knowledge, has a direct bearing on the scientist's aspiration for even more information about the meaning of the Universe (Blish).

Re-enactments of myth in science fiction take several forms. The simplest strives to deepen the emotional connotations of a story by permeating it with the reverberations of some great original, as C.S Lewis does successfully with the myth of the temptation of *Eve in Perelandra* (1943).

A less complex and yet popular strategy for mythology stories is to tell the myths from the time in which they happened, rationalizing them in the process. Such rationalization or revision takes place in *The Keltiad* series by Patricia Kennealy Morrison. The author blends science fiction and fantasy as it transposes the ancient Celtic world and its customs into a space-faring future. The premise behind the series is that a group of Keltic people left earth in 453 C.E in their spaceships and travelled to outer space to create a new kingdom, Kelta.

Three novels in *The Keltiad* series deal specifically with Arthur and are known as 'The Tales of Arthur' – *The Hawk's Gray Feather* (1990), *The Oak above the Kings* (1994) and *The Hedge of Mist* (1996). All three books provide an extensive treatment of Arthur's descendants and their history. Kennealy-Morrison's novels create an extensive family tree that continues beyond King Arthur, providing for his descendants to succeed him as rulers of the kingdom. By setting her novels in outer space, Kennealy-Morrison does not provide Arthur with a bloodline that connects him to the humans of twentieth century earth. Rather the 'Tales of Arthur' novels take place in the 21st century. Outer space is that which leaves the novels open for such extensive creation of descendants for Arthur, providing interesting possibilities for the legend, even if they are greatly removed from the traditional story.

While it is easy to disclaim this as disloyalty to tradition, Kennealy-Morrison in a way is truer to tradition than purists who simply translate or retell. MacKillop notes that myths have never been static, and the very idea of a right version of an excerpt of the narrative of human imaginative story is flawed (MacKillop 40). The revision's contemporary relevance and vitality is almost necessary. Her books help to illustrate this point. Though different from its likely origin, the Arthurian myth remained alive, and so more likely to be continued, at least in part than a static official version.

Dan Simmons also draws on literature and mythology in his works. In *Ilium* he draws on Greek legend and mythology, recreating *The Iliad* in a science fictional context, with people taking on the roles of Greek Gods and the war itself taking place on Mars. Three interweaving storylines (and timelines) interact in the telling of this story which is replete with elements like Artificial Intelligences and time travel. *Olympos* is its sequel. Both the *Ilium* and the *Olympos* are a form of literary science fiction, and rely heavily on intertextuality particularly Homer and Shakespeare as well as references to Marcel Proust's *A la Recherche du Temps Purdue* and Vladimir Nabokov's novel *Ada or Ardor: A Family Chronicle* (Wagner).

In *Olympos*, Simmons utilized the revisionist mythic possibilities afforded him by the *Iliad*. 'Further, a pastiche of *The Tempest* is what *Olympos*' earthbound story thread morphs into' (Wagner). At a deeper level, *Ilium* and *Olympos* lend their fiction the 'purely human art of storytelling' (Wagner) and the way that heroic myths and romances and tragedies have shaped and continue to shape civilization. Whether through his own revising of Homer's and Shakespeare's characters and sagas, or the obsession his cyborg Moravec have with ancient human literature, or through his depiction of the vacuity of the lives of his human characters, Simmons drives home the theme that art is important to who we are as anything else in history (Wagner).

One of the ways the past informs the present is through the mythic narratives that are passed down from generation to generation. Just as humanity evolves, so too do the myths, and *Olympos* offers an example of this. 'In revising and re-contextualizing the epics of Homer and Shakespeare,' Simmons seems keenly aware that in future generations, revisions of contemporary epics are revised (Wagner).

Mythology in *Star Trek*

Star Trek has existed for half a century, and has accumulated a vast body of work which results in a total of 738 hours of viewing time. From the original series through *Star Trek* enterprise, plot lines often directly or indirectly allude to religions and myths (Grech 23). Richard Lutz asserted

> The enduring popularity of *Star Trek* is due to the underlying mythology which binds fans together by virtue of their shared love of stories involving exploration, discovery, adventure and friendship that promote an egalitarian and peace loving society where technology and diversity are valued rather than feared and citizens work together for the greater good. Thus *Star Trek* offers a hopeful vision of the future and a template for our lives and our society that we can aspire to (Lutz 1).

However for an adequate analysis of the mythic qualities present in *Star Trek,* a close examination of the definition of myth has to be made. Blake Tyrell defined myth as

> [...] narratives with the power to move our psychic energies toward integration of self and of self with the cosmos. Myths define an image of the world within and without and relate us to it emotionally. Myths put in narrative from the unconscious assumptions that constitute the spirit of a culture. They can

inspire and direct those energies to monumental achievements of good or ill (Blake Tyrell 712).

Star Trek proffers many of the standard elements of mythic structure, fulfilling in the process the mythic functions of being mystical, cosmological, sociological and pedagogical. Campbell, in his seminal work of comparative mythology explores the theory that important myths from around the world which have survived for thousands of years all share a fundamental structure, which he calls the 'monomyth' (Campbell 23).

In laying out the monomyth, Campbell describes a number of stages or steps along this journey. The hero starts in the ordinary world and receives a call to enter an unusual world of strange powers and events. If the hero accepts the call, the hero must face tasks and trials and may have to face these trials either alone or he may have assistance. *Star Trek*'s epic hero is Captain James Kirk. He displays charisma, bravery and independence – becoming an icon of contemporary culture. Though Kirk stands for the ideal of independence, he is not complete on his own. The friendship he forged with Mr Spock – a Vulcan, parallels the model of the epic hero and his double. It is also a representation of the American myth, in which 'the most enduring and respected American classics revolve around the friendships of two males usually of two different races (Selley 94). This bond is consummated when Kirk and Spock mind-meld (Abrams). The Vulcan mind meld is a telepathic link between two individuals, allowing for the exchange of thoughts, thus in essence allowing the participants to become one mind (Landau). It is a psionic technique for 'synaptic pattern displacement' (Landau) and is employed only by Vulcans.

Though the friendship between Kirk and Spock forms the crux of the *Star Trek* mythos, the other members of the crew are also essential players in this epic text. Ellington and Critelli applied the Jungian symbols to the Enterprises' four senior officers who 'form a perfect quaternity of opposing personality types' – Kirk and Mc Coy embody the 'extroverted, intuitive, thinking type' while Scott and Spock are representations of 'introverted sensation, feeling type[s]. Ellington and Critelli contend that together these 'symbols of hope' create a 'model of effective functioning for personality as a whole' (Ellington and Critelli 302). The viewers are able to identify themselves with these representations, juxtaposing the mythic narratives in *Star Trek* to the emotional needs of the viewer (Zimmel 23).

Following this tradition, *Star Trek* provides a vision of friendship between two men of different races and different attitudes. This relationship is a central point of various episodes. A brilliant representation is rendered in 'Amok Time' (Pevney). On Spock's home planet Vulcan, Mr Spock is forced to fight a legal battle against Kirk to be allowed to marry according to cultural traditions. Kirk gives his life but Spock's desire for his future wife is gone: 'I will do neither [live and prosper], for I have killed my captain and my friend' (Pevney).

The mission of the *Enterprise* undoubtedly resembles the adventures of Homer's *The Odyssey* - old Greek myths that are revolutionized in a science fiction format. Zimmel postulates that the voyage is one of the 'most important similarities to classical myths: protagonists journeying through formerly unknown countries, braving all kinds of danger and trials to test their capacities' (Zimmel 24). The mission of the *Enterprise* is similar – it is sent to explore unchartered space relying on their 'individual skills in many difficult situations' (Zimmel 25).

The Romulans and the Vulcans are modern representations of the founding of the Roman Empire. In the Roman myth Romulus and Remus are the twin brothers that are the focus of the Rome creation myth. In the myth, Remus and Romulus are raised by a wolf and go on to found the Roman Empire. Remus and Romulus disagree on the location for the new empire. The dispute elevate and gets physical leading to the death of Remus by

Romulus. After Remus's death Romulus goes on to found Rome. Similarly, Romulans and Vulcans share a common ancestry. They had both lived on the Vulcan planet and were one race. During the time of 'Awakening' – the violent and warlike Vulcan race switched to more peaceful ways (Dawson). The change was brought on by a Vulcan called Surak who preached a philosophy of using logic to control emotions, as well as total pacifism. He was effective in spreading his teaching across Vulcan. However, a sizeable group of Vulcans resisted his efforts, leading to a war using weapons weapons of mass destruction, such as atomic bombs. Ultimately, the warlike Vulcans left the planet becoming the ancestors of the Romulans. They began to formulate the Romulan Empire which parallels the Roman Empire.

The Greek creation myth is also alluded to in the Klingon religion. A version of this creation myth is told during the traditional Klingon wedding ceremony in the episode 'You are cordially invited' (Livingstone).

> With fire and steel did the gods forge the Klingon heart. So fiercely did it beat, so loved was the sound that the gods cried out, 'On this day we have brought forth the strongest heart in all the heavens. None can stand before it without trembling at its strength.' But then the Klingon heart weakened, its steady

rhythm faltered and the gods said, 'Why have you weaken so? We have made you the strongest in all creation' (Livingstone).

When the heart expressed the solitude of being left alone, the gods knew that they had erred, so they forged another heart, stronger to beat together, destroying the gods who created them and turning the heavens to ashes.

Writer Ursula K. Le Guin once noted that 'science fiction is the mythology of the modern world' (Le Guin www.ursula.kleguin.com). She contends that it is an arrangement of old motifs utilized again, effectively portrayed in the connection between science fiction and Judeo-Christian mythology. The stories range from search for paradise to the eternal struggle between good and evil which forms the mythical perspective. Further, Zimmel insists that in *Star Trek* the 'fandom itself seems to have a nearly religious devotion to the series' (Zimmel 26). Jewwett and Lawrence go on to compare the immense magazine literature that evolved around the show – to 'acrophyal literature in the biblical tradition' (Jewwet and Lawrence 50). This kind of writing answers essentially theological questions, 'simplifying and illustrating a faith' (Isaiah 11.6).

The Mythology of *Stargate*

Stargate is a military science fiction franchise, initially conceived by Roland Emmerich and Dean Devlin, based on the idea of an alien wormhole, that is, an Einsten-Rosen bridge (the Stargate) that enables nearly instantaneous travel across the cosmos. Stargate productions centre on the premise of a Stargates that enable instantaneous transportation to another devices located astronomical distances away (Emmerich).

One of the grounding elements of the series overall story arc and a key aspect of its appeal is Egyptian mythology. Central to the framework of the *Stargate* universe is that the cannibalistic, warlike Goa'uld poses as the mythological gods of human cultures. A direct cultural link between Teotihuacan and the Aztecs, which does not exist in the real world, was also part of the Stargate framework, with an artefact found on the planet Orban referring to an Aztec goddess named Chalchuhthau (Turner).

Aztec mythology has a pantheon of bloodthirsty deities that can readily be adapted to fit the *Stargate* framework. In mythology, the god Huitzilopochtli, who found the Aztec Empire was born with all the knowledge of his Mayan mother, Coatlicue. Painted images of Coatlicue bear a remarkably convenient resemblamce to a principle antagonist in *Stargate*: the Goa'uld queen Hathor (Turner).

To fulfill the requirements of using technologized mythology for the *Stargate* book *Exogenesis* (Whitelaw and Christensen) requires the seamless linking of real world mythology to *Stargate* mythology. In *Stargate*, the indifferent protagonists – and sometimes antagonists – are the Ancients, who bear the names and often, the attributes of the gods, demigods and heroes of ancient Greece and Rome. In Earth mythology, Plato described Atlas the first king of Atlantis (Raphael). In the *Stargate* framework, however, Moros was the Ancient Leader of Atlantis ten thousand years ago (Turner). Extrapolation from this in *Exogenesis* (Whitelaw and Christensen), allows for the possibility of an antagonistic relationship between the Ancients Atlas and Moros, the latter of whom might conceivably have displaced Atlas and Moros from a position of power. This contradiction is resolved in *Exogenesis* by suggesting that Atlas and Janus were aligned while Moros' objectives to the temporally-enhanced exogenesis machine were personal (Whitelaw and Christensen).

Using a terraforming device in a compressed time frame of days or weeks by Ancients who were known to have recreated all life in the Milky Way Galaxy also connects to the creation myth in Genesis while there are several references to Biblical Mythology in *Exogenesis.* Using the creation god, Ea from Gilgamesh as an ancient character sidesteps the potentially delicate marketing issue of using the Judeo- Christian God as a character in *Stargate* (Whitelaw and Christensen).

Ea, god of primordial waters, is credited with creating Earth, and sometimes also credited with creating mankind through genetic manipulation, and of warning the Sumerian Noah of a great flood sent by the other gods to destroy mankind. In *Exogenesis,* Ea is an Ancient terraforming engineer who, despite her actions and grief, regretted her behaviour (Whitelaw and Christensen).

Nabu's role in *Exogenesis* also mirrors that of his mythological counterpart. In Babylonian mythology he is Ea's grandson, the god of wisdom and writing who rides a winged dragon. Regarded as a demi-god, he has the power to alter the length of a human life by writing their destiny on a stone tablet. In *Exogenesis,* Nabu is a half human half Ancient who rides a dragon-like Dart, a scholar who interceded with the destiny of the genetically altered humans (Whitelaw and Christensen).

Discussion

Myths are everywhere, tingeing the blandest discourse with dire resonance making the mildest encounter a drama. Disch sustains that even Freud, Levi-Strauss and Barthes sustain that mythology, in a very broad sense, embraces the whole realm of the cultivated and the civilized shaped by the hand and mind of men, which for most of us includes everything in sight. Myths are especially present in literature, and even more so in science fiction. The reasons for this are tangible. Myths aim at maximizing meaning, at compressing truth to the highest density that the mind can assimilate. To attain such compression, myths make free use of the resources of the unconscious mind, that alternate world where magic still works and metamorphoses are an everyday occurrence. In fact, Disch claims that science fiction has been trafficking in magic and mythology since it first came into existence. Mary Shelley's *Frankenstein,* is subtitled *A Modern Prometheus,* and the horror show monster whose image continues to be emblematic of the genre is probably the descendant of 'Gorgons and Hydras, and Chimaeras dire' (Disch 21).

As mythmakers, science fiction writers have a double task, the first aspect of which is to make humanely relevant the formidable landscapes of the atomic era. Quite often, in searching for a place to install one of the latter science fiction figures, the safe author discovers that the new figure 'corresponds very neatly with one already there' (Disch 22), though this might prove difficult at times.

The second task of science fiction writers as mythmakers is simply the custodial work of keeping the inherited body of myths alive. Every myth is the creature, originally of a poet, and it remains a vital presence in our culture only so long as it speaks to us with the living breath of living art. The names might be changed, the scenery altered, but the basic patterns are fixed.

Science fiction is an exciting genre of literature. Science fiction and fantasy are by their very natures in a position to become the contemporary mythology. Mythology is grounded in metaphor and Clayton claims that metaphors and symbols of mythology that feed the human spirit flow out of the story - the saga. And at the heart of such story, the midst of a dangerous and often unknown setting, human ingenuity and courage and faithfulness are put to the test. The mythic story is also spatial – it often involves unknown, perhaps terrifying terrain. The outer space, thus provides an ideal setting for the contemporary myth (Clayton)

The genius of science fiction lies in the fact that the reader is caught up in a great story in the midst of a strange world, long way from earth and from our own time. With the turn of a phrase or a twist of perspective, suddenly the reading is given the chance to look introspectively. A good myth does this as well. Science fiction as myth can be a window into our own souls, a way of sounding the depths of the human psyche. And science fiction is a popular way of spinning a contemporary entertaining story even while exploring serious ideas.

Both myth and science attempt to provide an overview of existence by bridging inner with outer reality. Myth attempts to project inner reality (conscious desires, archetypal patterns) in a metaphor for outer reality, while science aims to illuminate inner reality through the study of outer, empirical forms. Sutton and Sutton contend that a body of myth forms an autonomous universe which stands in metaphoric relation to the actual world. Scientific hypotheses also form a universe, a universe which is not identical to objective reality but representative of man's understanding of it. Thus the question of validation or disproof is irrelevant to myth since the relation of myth to reality is analogical, but it is paramount for science because the worth of a scientific hypothesis is entirely dependent on the accuracy of its relationship to objective reality (Sutton and Sutton 232).

Gilkey claims that before the advent of the scientific mode the only means by which man could relate to the universe was through the mythopoeic mode. His acceptance of the narratives of gods and heroes as the meaning of the world served as an affirmation of space, of time, of natural occurrence and of a historical event (Gilkey 286).

Prescientific man viewed everything outside himself as 'other' and to a large degree unknowable. For him myth served as the vehicle for his relationship with the 'other'. As the scientific or technological mode developed, man's orientation moved away from universal concepts to a more specialized focus on the individual empirical data (Gilkey 286). Historically this shift resulted in the sharp distinction between the two modes of thought, with the scientific recognized as the means to knowledge and the mythopoeic disenfranchised and relegated to the role of plaything for poets. Sutton and Sutton insist that now readers are in a position to move beyond this convenient dissection of thought, for the mythopoeic and the scientific modes in their matured states are now disintegrating. The present situation must be viewed in relation to a transcendent order of some description. For early mythopoeic man, this transcendent order was the cosmos with its god, heroes, planets and other inexplicable phenomena (Sutton and Sutton 236). In the area of scientific myth, the transcendent referent can no longer be the cosmos, some scientific research has shown that it is empirically knowable and as a consequence it is no longer entirely transcendent. As a referent, modern myth especially science fiction, replaces the cosmos with the concept of space. This archetype, so vital to humanity, has been expressed throughout human history in various forms, but it is characteristic of our time that it should take the form of a technological construction in order to avoid mythological personification (Ellul 221).

Space-time lend science fiction an infinite, unknown extension which lends a grandeur to whatever actions are undertaken in it. Unlike a scientific hypothesis, a science fiction story is not formulated primarily to advance technological knowledge, but rather operates on a visionary mythopoeic level. Thus science fiction is a self-conscious form of myth in which man intentionally mythologizes scientific narrative.

Bibliography

Cinematography

"Amok Time." Dir. Joseph Pevney. *Star Trek: The Original Series.* Paramount. September 1967.

"Awakening." Dir. Roxann Dawson. *Star Trek: Voyager.* Paramount. November 2004.

"Sarek." Dir. Les Landau. *Star Trek:The Next Generation.* Paramount. May 1990.

"Stargate." Dir. Roland Emmerich. Metro-Goldwyn-Mayer. October 1994.

"Stargate: Atlantis." Dir. Brad Turner. Sony Pictures Television. 2005-2006.

"Stargate: SG-1." Dir. Brad Turner. Sony Pictures Televesion. July 1997.

"Star Trek: The Future Begins." Dir. J.J. Abrams. Paramount. May 2009.

"You are Cordially Invited." Dir. David Livingstone. *Star Trek: Deep Space Nine.* Paramount. November 1997.

Primary and Secondary Texts

Bascom, William. "The Forms of Folklore: Prose Narratives." *Sacred Narrative: Readings in the Theory of*

Myth. Ed. Alan Dundes. Berkeley: University of California Press, 1984: 5-29.

Blake Tyrell, William. *Journal of Popular Culture.* X/4:712.

Blish, James. "Mythology." *The Encyclopedia of Science Fiction.* April 2015

www.sf-encyclopedia.com/entrymythology Accessed 20 January 2016.

Campbell, Joseph. *The Hero with a Thousand Faces.* Princeton: Princeton University Press, 1968.

Clayton, Ron L. "Science Fiction as Contemporary Mythology."

      www.wyrdwanderers.wordpress.com Accessed 6 February 2016.

Disch, Thomas. *Mythology and Science Fiction.* USA: The University of Michigan Press.

Ellington, Jane Elizabeth and Critelli, Joseph W. "Analysis of a Modern Myth: The Star Trek Series." *Extrapolation.* 24.3. 1983:241 - 250.

Ellul, Jacques. *The Technological Society.* New York. 1965: XIX.

Fava, Anna. "Science Fiction – Mytholgoy of the Future." *Think.* Malta: University of Malta.

Gilkey, Langdon. "Modern Myth-Making and the Possibilities of Twentieth Century Theology." *Theology of Renewal.* Montreal. I,1968: 286.

Grech, Victor. "The Pygmalion-Galatea Myth in Relation to Simulation Scenarios in Star Trek." *Xjenza Online – Journal of Malta Chamber of Scientists.* 2.3. 2013: 23 - 28.

Hodge, James,L. "New Bottles – Old Wine: The Persistence of the Heroic Figure in the Mythology of

Television Science Fiction and Fantasy." *The Journal of Popular Culture.* 21.4. 1988: 37-48.

Jewett, Robert and Lawrence, John Shelton. *The American Monomyth.* New York: Doubleday, 1977.

Kennealy-Morrison, Patricia. *The Hawk's Gray Feather.* United Kingdom: Roc. 1990.

Kennealy-Morrison, Patricia. *The Hedge of Mist.* United Kingdom: Roc. 1996.

Kennealy-Morrison, Patricia. *The Keltiad Series.* United Kingdom: Roc. 1988.

Kennealy-Morrison, Patricia. *The Oak above the Kings.* United Kingdom: Roc. 1994.

Koven, Mikel J. "Folklore Studies and Popular Film and Television: A Necessary Critical Survey." *Journal of American Folklore.* 116.460. 2003: 176-195.

Le Guin, Ursula K. "Plausibility Revisited: What Happens and What Didn't."
www.ursulakleguin.com Accessed 27 January 2016.

Lewis, C.S. *Perelandra.* United Kingdom: The Bodley Head, 1943.

Lutz, Richard. "Social Cohesiveness." *Human Rights Coalition.* Australia.

Mac Killop James. *Fionn Mac Cumhaill: Celtic Myth in English Literature.* New York: Syracuse University Press, 1986.

New Internation Version. [Colorado Springs]: Biblica, 2011.
www.biblegateway.com Accessed 27 January 2016.

Raphael. "Plato from the School of Athens." *Portal Philosophy.* 1509.

Selley, April. *Journal of Popular Culture.* 20/1 (Summer). 1986: 94.

Simmons, Dan. *Ilium.* United States: Harper Collins, 2003.

Simmons, Dan. *Olympos.* United States: Harper Collins, 2005.

Singer, Irving. "Introduction: Philosophical Dimensions of Myth and Cinema." *Cinematic Mythmaking: Philosophy in Film.* Massachusetts: MIT Press Books, 2008.

Sutton, Thomas C. and Sutton, Marilyn. "Science Fiction as Mythology." *Western Folklore.* 28.4. (October), 1969: 230-237.

Wagner, T.M. *SF Reviews Net.*
        www.sfreviews.net Accessed 20 January 2016.

Whitelaw, Sonny and Christensen, Elizabeth. *Exogenesis.* USA: Fandemonium, 2006.

Zimmel, Daniel. "Just a Television Show? The Myth of Star Trek." *Online Book.* 1998.

Chapter 9. Sentience in Science Fiction[4] – Mariella Scerri and Victor Grech

The Oxford Dictionaries define sentience as the ability to perceive or feel things. Bortolotti and Harris emphasise the distinction between the capacity to have experiences and react appropriately to external stimuli (sentience) and the additional capacity to be aware of oneself as a distinct individual whose existence began sometime in the past and will extend into the future (self-consciousness). The authors contend reactive behaviour without intentionality is not 'sentience' as it does not involve phenomenal consciousness. It is merely the capacity to react to external stimuli. Plants and computers have this property without being aware of the qualitative aspects of the stimuli they react to. Having phenomenal conscious experiences requires the awareness of some qualitative aspects (or qualia) of the experiences, for instances the brightness of a colour one visually perceives (Dennet).

Another characterization of sentience is the capacity to feel emotions, such as pain or pleasure. While plants and computers react to external stimuli, they do not feel emotions. This concept is central to the

---

[4] This was originally published in SFRA as Scerri, Mariella and Grech, Victor. "Sentience in Science Fiction." *Science Fiction Research Association.* 315 (2016): 14-18.

philosophy of animal rights, because sentience is necessary for the ability to suffer, and thus is held to confer certain rights. Ned Block asserts that 'fundamentally different physical realization from us per se is not a ground of rational belief in lack of consciousness' (Block 392). Marc Bekoff believes that people are not exceptional or alone in the arena of sentience. He insists we need to abandon the anthropocentric view that only big-brained animals such as ourselves, non-human great apes, elephants and cetaceans have sufficient mental capacities for complex forms of sentience and consciousness.

In science fiction, an alien, android, robot, hologram or computer described as 'sentient' is usually treated in the same way as a human. Foremost among these properties is human level intelligence (that is sapience) but sentient characters also typically display desire, will, consciousness, ethic, personality, insight and humour. Sentience is being used in this context to describe an essential human property that brings all these other qualities with it. The words 'sapience', 'self-awareness' and 'consciousness' are used in similar ways and sometimes interchangeably in science fiction.

Science fiction has explored several other forms of consciousness besides that of humanity, and how such minds might perceive and function. In "The Pinocchio Syndrome and the Prosthetic Impulse in

Science Fiction," Grech (2012) opines that three components constitute the mental and psychological aspects that define man; 'the desire to acquire 'qualia', the expression of intentionality; and an application of an Abraham Maslow-type motivational pyramid, with a desire for self-actualisation that embraces the desire to attain humanity.' These three facets, Grech notes are demonstrated through the character Data in *Star Trek*. Those who meet Commander Data will reasonably be sure that he is conscious. However finding out that he is not human does not cancel that ground for rational belief. Block argues 'the root of the epistemic problem is that the example of consciousness on which it is inevitably based is us. But how can science based on us generalize to creatures that do not share our physical properties?' (Block 295).

Block furthermore claims that naturalism asserts that the default position is that Commander Data, being an artificial construct, is not conscious. On the other hand, disjunctivism allows that if Commander Data is conscious, shared phenomenality is constituted by the property of having Commander Data's electronic or electro-chemical realization of our functional state.

Such debates can provide a basis and a framework for the issues of sentience and non-sentience that come up in Science Fiction narratives. This paper aims to analyse how sentience is treated in Mary Shelley's *Frankenstien,* in the *Star Doc* Series, notably the first book in the series, *Star Doc* and in specific episodes in *Star Trek.*

Frankenstein

Frankenstein is said to be the first Science Fiction novel (Wilson Aldiss). The trope of sentience was mooted with regard to Victor Frankenstein's monster. The monster's sapience is significant throughout the book with several interjections by the monster himself about feelings of rejection and loneliness. On the other hand, Frankenstein's ambivalence toward his creation reinforces the frankly callous scepticism he held toward the monster as a sentient life form. Indeed, the monster remains unnamed and is instead referred to as 'monster,' 'creature,' 'demon,' 'devil,' 'fiend,' 'witch' and 'it.'

Fast forward in time, readers of science fiction encounter the same ambivalence in the treatment of sentience in science fiction narratives.

Sentience in Viehl's Star Doc

Doctor Cherijo Grey Veil is a doctor and surgeon who accepts a position as a physician at Kevarzanga-2's Free Clinic. Her surgical expertise is desperately needed on a frontier world with over two hundred sentient species, and her understanding of alien physiology is a consequence of a keen intelligence and an eidetic memory. But there is a hidden truth behind her expertise. Dr Cherijo is a genetically enhanced clone, an experiment conducted by her father who is the archetypical cold, calculating and ruthless physician. After a series of ten unsuccessful attempts, Dr. Cherijo proves to be the only successful outcome of these experiments. She is superhuman with a superior capacity for learning and an enhanced immune system which transcends that of mundane humanity.

The denial of this individual's sentience reaches its denouement with a rigorous four day trial, and the decision for subsequent deportment of the protagonist to Terra (planet Earth), because it has been proven that her existence, the result of Joseph Grey Veil's experimentation and his violation of the "Genetic Exclusivity Act" breaks 'Section nine, paragraphs two through four of the League's Treatise which prohibits such experiments.

Her only minimal chance for an appeal, as suggested by Dr Mayer, chief medical officer, is to petition to the ruling council with an emergency request to be declared a sentient being. The protagonist is 'a clone-created, modified, trained and being observed during an extended experiment. You are not classified as human or sentient. You are Joseph Grey Veil's property.' Being genetically enhanced during embryonic development, she is deemed unclassifiable as the "Genetic Exclusivity Act" has been breached.

The best reason for Cherijo to be declared sentient is given by nurse Ecla. She claims that non-sentient life forms do not have the ability to understand the meaning of death. Nonetheless, during an epidemic, Dr Grey was seen many times 'holding a dead child in her arms, and praying to her God for that lost little soul'. What is even more bigoted in Dr Grey Veil's trial is the criteria for which she did not meet and thus denied sentient status: she had not been conceived, gestated or delivered by natural or legally sanctioned methods; was in possession of 'enhancement deliberately bred by experimentation'; and never been allowed to live freely. These three main criteria move away from the epistemology of consciousness per se. However, Block debates the role of functional similarity in providing evidence that others are like us in intrinsic physical respects, and that is the ground for our belief in other minds.

Throughout the *Star Doc* series we encounter other life forms with similar issues related to sentience. The sentience status of a Chakacat called Alunthri, a human sized cat with human-equivalent intellectual abilities and language skills is raised and debated in *Star Doc*. Chakacats 'once captured and trained' are sold as domesticates […] there is some controversy about their classification. Effort by Council petition to have them recognized as sentient life forms have been consistently denied' (Viehl 80).

The deliberate stance taken by Dr Grey Veil is 'Alunthri, I couldn't treat you like a domesticated companion. In my eyes you are sentient' (Viehl 185) parallels Block's arguments in favour of sentience. This occurs when Alunthri seeks her assistance to transfer him under her ownership. Without deed, under the terms of the current colonial charter, he would be shipped back to his home world Chakara and resold. He specifically asked for her ownership because he knew that Veil would give him this freedom. The working definition of sentience comes into full force here where the ability to feel, perceive or to experience subjectivity is most palpable.

Sentience in Star Trek

Similar issues on sentience also arise in Star Trek. In 2365, Phillipa Louvois of the Judge Advocate General's Office held a hearing in which she decided that Data was not the property of Starfleet. During the hearing the question of an android's sentience came up but there was no formal, legal resolution on the matter (TNG: "The Measure of Man"). Despite a lack of official acknowledgement, Data thought himself to be sentient and many others agreed. (TNG,

"The Offspring," "The Most Toys,") so much so that as of 2371, Data was considered the only sentient artificial lifeform in Federation Society (VOY, "Prototype").

From time to time other non-android life forms or artificial intelligences have been considered sentient as well. In the episode "Warhead" (Kretchmer), a weapon was so sophisticated that it was considered sentient. Holograms have also been referred to as both artificial lifeforms and 'sentient.' One such sentient hologram was created on the USS Enterprise – D in 2365, when Lieutenant Commander Geordie La Forge requested that the holodeck create an opponent worthy of Data in a Sherlock Holmes – style mystery. The ship's computer produced a sentient version of James Moriarty, Holmes' nemesis.

A legal case related to holographic sentience arose with the *Voyager* Doctor when he attempted to publish a holonovel entitled "Photons Be Free," but it was appropriated and released without his permission by his publisher. The legal issue revolved around whether the doctor was an 'artist' within the meaning of the laws that granted rights to control the dissemination of intellectual property. The ruling was narrow in that the definition of artist in that single law was extended to a hologram, but it was an important step on the path toward granting full legal status to a hologram as a sentient entity (VOY, "Author, Author").

Discussion

The treatment of sentience in science fiction narratives has been a cause of ambivalence toward acceptance of sentient non-human life forms and their quest for human rights, with both the legal and ethical implications that this may bring.

In *Star Doc* and *Star Trek*, the same hesitancy to accept sentient life forms is encountered. Both Doctor Cherijo and Data are artificial life forms. Doctor Cherijo is the result of a successful laboratory experiment carried out by her father, while Commander Data is an android possessing excessive rationalism and incapable of conveying emotions. The implicit notion of being seen as the 'Other' is explicit throughout various incidents culminating in the trials they both had to undergo. These implications seem to suggest that while science fiction narratives acknowledge sapience and sentience in other life forms, these same books resist giving the prescribed rights, both ethical and legal, which accompany human beings.

The recent film *Ex-Machina* written and directed by Alex Garland implies the same resistance in treating man-made life forms. *Ex Machina* takes us into the not too distant future where a genius billionaire has created the world's first fully sentient artificial intelligence, in the beguiling female form of Ava. He invites a low level employee to his remote laboratory home to apply the Turing Test to his creation. The film tried to marry the juxtaposition inherent in the central idea that she is man-made, but that Caleb is there to wonder if intelligence is necessarily human, and whether she has human intelligence. Ava is not fully robotic nor fully skinned or human, thus the viewers are constantly reminded that she is still a machine.

Issues on sentience in these narratives lend themselves to contemporary debates such as stem cell research, personhood and sentience. Bartolotti and Harris in their paper, 'Stem Cell Research, Personhood and Sentience' claim that in ordinary language we identify persons with human beings but the notion of a person is not co-extensive with the notion of a human being. More specifically, whereas an individual counts as a human being if it belongs to the species *Homo Sapiens*, it counts as a person not by virtue of species membership, but of the capacities it possesses. Bortoletti and Harris contend that empirical studies rule out that human embryos and foetuses are persons, as they do not satisfy the requirements for personhood i.e. rationality and self-consciousness. The conclusion is that it is immoral to prevent the development of an embryo because the embryo has the potential to become a person. This relies on the assumption that one should treat a potential person as one treats a person. However, there are direct moral obligations toward persons by virtue of their interests in their own well-being. Is it justified to grant the same moral status to early embryos that have no interests in their own well-being?

On the other hand, according to the principle of human dignity, in a formulation that can be found in Kant (1785), human life should never be thought of merely as a means but always also as an end. Inspired by Kant's formulations, some might argue that human embryos cannot be just treated as a means to further research as this would violate the principle of human dignity. Steinbock (1997) and Roberston (1995) shed light on another important viewpoint. They claim that human embryos occupy that space in between fully-fledged persons with rights and interests and insentient beings with no symbolic value.

Personhood and sentience are often argued for their moral significance. In both science fiction narratives and in real life, what defines life forms as sentient falls in a grey area lending itself to the numerous debates on the issues of sentience.

References
Cinematography

"Author, Author." Dir. David Livingstone. *Star Trek: Voyager.* Paramount. April, 2001.

"Ex Machina." Dir. Alex Garland. Universal Pictures International. January, 2015.

"Prototype." Dir. Jonathan Frakes. *Star Trek: Voyager.* Paramount. April, 2001.

"The Measure of Man." Dir. Robert Scheerer. *Star Trek: The Next Generation.* Paramount. February, 1989.

"The Most Toys." Dir. Tim Bond. *Star Trek: The Next Generation.* Paramount. May, 1990.

"The Offspring." Dir. Jonathan Frakes. *Star Trek: The Next Generation.* Paramount. March, 1990.

"Warhead." Dir. John Kretchmer. *Star Trek: Voyager.* Paramount. May, 1999.

Bibliography

Bekoff, Marc. "The Hearts and Minds of Animals: A discussion with Dr. Marc Bekoff." *Forbes.* Web. May, 2012.

Block, Ned. "The Harder Problem of Consciousness." *The Journal of Philosophy."* XCIC, 8. August, 2002:391-425.

Bortolotti, Lisa and Harris, John. "Stem Cell research, personhood and sentience." *Ethics, Law and Moral Philosophy of Reproductive Biomedicine.* 1(1) March 2005: 68-75.

Dennet, Daniel. "Are we explaining consciousness yet?" *Cognition.* 79(1-2) April 2001: 221-237

Grech, Victor. "The Pinocchio Syndrome and the Prosthetic Impulse in Science Fiction." *The New York Review of Science Fiction.* 284; 24(8). April 2012:11-16.

Kant, Immanuel. *The Metaphysics of Morals.* Gregor M. (transl.). New York: Cambridge University Press, 1996.

Robertson, John. "Symbolic Issues in Embryo Research." *Hasting Center Report.* 1995. 25:37-38.

"Sentience." Def. *Oxford Advanced Learner's Dictionary.* Oxford Dictionaries, n.d. Web: 22 July, 2015.

Shelley, Mary. *Frankenstein, or The Modern Prometheus.* Project Gutenburg Ebook: June, 2008.

Steinbock, Bonnie. *Life Before Birth: The Moral and Legal Status of Embryos and Fetuses.* Oxford: Oxford University Press, 1992.

Viehl, S.L. *Star Doc.* New York: Roc Books, 2000.

Wilson Aldiss, Brian. *Billion Year Spree: The True History of Science Fiction.* New York: Doubleday Publishers, 1973.

Note about the Authors

Ms Mariella Scerri graduated as a B.Sc (Hons) Nurse from the University of Malta in 1999. She has worked as a senior nurse within the cardiology department at Mater Dei  Hospital, Malta for the past sixteen years. She has also completed a B.A (Hons) Degree in English and Comparative Literature with Goldsmiths College, University of London and a post graduate course in certified education in English with the University of Malta. She is now a teacher of English at St Clare College Secondary School Pembroke. Ms Scerri is currently reading for a M.A in English at the University of Malta. She is a member of HUMS – a programme for Humanities, Medicine and Sciences Programme at the same university. She is also a Lecturer at the faculty of Medical Emergency Education of the Instute of

Technology, Humanities, the Arts, Medicine and Science as well as the executice secretary of the faculty of Medical Emergency Education. (www.imee-edu.com)

She also holds the position of Director at the Institute of Medical Emergency Education. She is married and a proud mother of two.
Ms Scerri finds creative writing, research and editing an interesting past time.

Prof. Victor Grech graduated MD from the University of Malta Medical School in 1988. He specialized in paediatrics and took up paediatric cardiology at The Hospital for Sick Children at Great  Ormond Street in London. While there, he commenced a Ph.D. entitled 'Congenital Heart Disease in Malta', and this was completed in 1998. His appointment with the Maltese Department of Health at Mater Dei Hospital, is as a consultant paediatrician with a special interest in paediatric cardiology.

Prof. Grech has published extensively not only in paediatric cardiology but also in general paediatrics, other aspects of medicine and in the humanities, particularly in the field of science fiction. He completed a second Ph.D. with the English Department at the University of Malta in 2011 entitled Infertility in Science Fiction. Prof. Grech is also the creator and editor-in-chief of the journal Images in Paediatric Cardiology (www.impaedcard.com), and is currently the editor of the Malta Medical Journal (http://www.um.edu.mt/umms/mmj). Current endeavors include co-chairing the Humanities, Medicine and Sciences Programme at the University of Malta and a dissertation involving male: female ratios at birth. Prof. Grech is also a Director and Master of Studies at the Institute of Technology, Humanities, the Arts, Medicine and Science (www.ithams.com) and a senior lecturer at the faculty of Medical Education. (www.mime.ithams.com)

Prof. Grech lives in Pembroke, Malta with his wife, two children and three Siamese cats, and finds painting Maltese landscapes and seascapes a particularly relaxing pastime. Some of his work can be found at www.maltaimpressions.com

Dr David John Zammit is specialised in computer studies and runs his own company Shadow Services Group. In 2008 he decided to do a PhD in Computers and Science Fiction, with his dissertation focusing on  "Science Fiction, Computers and Robotics." He completed a second Doctorate of Science in 2016. Dr David Zammit is the Director and Head of the Institute of Technology, Humanities, the Arts, Medicine and Science (www.ithams.com) and the Head of Academia of the Faculty of Medical Emergency Education (www.imee-edu.com). He is also the Commander in Chief, International of the Special Rescue Group - St Lazarus Corps (www.srg-int.org), and also the President of Femeraid International - Malta.

He has written a number of books, not only in International Trauma but also in First Aid, Basic Life Support and Emergency Rescue.

Since 1988, Dr. Zammit has served as a senior Manager, General Manager or director for multiple companies or Institutes.

David was interested in First Aid from a young age, joining the St. Join Ambulance as a cadet in 1975 when he was twelve years old. He became an instructor in First Aid in 1979. In 1981 he was promoted to Divisional Officer and later that year he was promoted to Divisional superintendent. He left St. John Ambulance in 1984 to further his studies, but in 1991 he started the Special Rescue Group – St. Lazarus Corps in Malta. The Group started to expend with its First International branch in 2004, today having a presence in six different countries. In 2013 he became Director and Head of Academia at the Institute of Medical Emergency Education, as well as serving as a senior lecturer in Emergency Medicine and Trauma International Life Support. He was appointed as the Head of the Institute of Technology, Humanities, the Arts, Medicine and Science in January of 2016

In May of 2015 he was appointed Chairman for the board of Health and Safety Certification standards.

Dr Zammit lives in Pembroke, Malta with his wife and two children. In his free time, he enjoys oil painting especially Maltese landscapes, reading and computer games. He is also a Knight Commander of the Hospitaller Order of St Lazarus of Jerusalem and a Knight Commander of Malta.

AFTERWORD

Afterword

Michael Dirda, a book critic and winner for the Pulitzer Prize in 1993 once said, 'Science Fiction is after all, the art of extrapolation.' The collection of papers presented in this book is an affirmation of such a statement. All the papers successfully explored various intersections between the humanities and the sciences in science fiction novels, films and television series. The need to document this wide and impressive array of research has led to the publication of this book. This is the second book in this series; the first book *Star Trek: Interdisciplinary Perspectives in Theory and Practice* was published in 2015. The timely publication of this book coincides with the 50th Anniversary from the launch of *Star Trek: The Original Series.*

Preparations are also underway for an upcoming *Star Trek* Symposium to be held in Malta in July 2016. The editors intend to publish the proceedings of this coming event as well. When all three books are published they will yield an interesting book series for both academics, science fiction readers and *Star Trek* fans alike.

'With Science Fiction there's endless possibilities' (Anna Torv) and with this befitting ending, the editors hope that the readers of this book have delved into the wide ranging possibilities offered as much as they have enjoyed editing and creating this book.

Further information on these symposia is found on:
www.scifisymposium.com
www.startreksymposium.com
www.scifi-malta.com

www.facebook.com/StarTrekSymposium
www.facebook.com/scifisymposium

Paintings
The paintings used in the front cover of the book are the artistic work of Dr. Zammit.

Sci fi Malta logo and design of the cover are the work of Dr. D Zammit.

www.ingramcontent.com/pod-product-compliance
Lightning Source LLC
Chambersburg PA
CBHW070228190526
45169CB00001B/118